Gesellschaft – Technik – Umwelt
Neue Folge

herausgegeben vom Institut für Technikfolgenabschätzung
und Systemanalyse (ITAS) am KIT Karlsruhe und
Prof. Dr. Armin Grunwald

Band 21

Peter Hocke | Sophie Kuppler
Ulrich Smeddinck | Thomas Hassel [Hrsg.]

Technical Monitoring and Long-Term Governance of Nuclear Waste

Die Deutsche Nationalbibliothek verzeichnet diese Publikation in der Deutschen Nationalbibliografie; detaillierte bibliografische Daten sind im Internet über http://dnb.d-nb.de abrufbar.

The Deutsche Nationalbibliothek lists this publication in the Deutsche Nationalbibliografie; detailed bibliographic data are available on the Internet at http://dnb.d-nb.de

ISBN 978-3-8487-4402-2 (Print)
978-3-8452-8659-4 (ePDF)

British Library Cataloguing-in-Publication Data
A catalogue record for this book is available from the British Library.

ISBN 978-3-8487-4402-2 (Print)
978-3-8452-8659-4 (ePDF)

Library of Congress Cataloging-in-Publication Data
Hocke, Peter | Kuppler, Sophie | Hassel, Thomas | Smeddinck, Ulrich
Technical Monitoring and Long-Term Governance of Nuclear Waste
Peter Hocke | Sophie Kuppler | Thomas Hassel | Ulrich Smeddinck (Eds.)
160 pp.
Includes bibliographic references.

ISBN 978-3-8487-4402-2 (Print)
978-3-8452-8659-4 (ePDF)

Onlineversion
Nomos eLibrary

edition sigma in der Nomos Verlagsgesellschaft

1. Auflage 2022

Inhalt

Sophie Kuppler, Peter Hocke, Ulrich Smeddinck, Thomas Hassel

Technical monitoring and long-term governance – an introduction

1 Overview

If high-level nuclear waste is disposed in a deep geological repository with the option of retrievability, it is necessary to monitor the development of the waste underground. Without the information gathered during monitoring, it would be impossible to know whether the state of the waste and the repository is as planned or whether an unwanted development has occurred. If the data collected deviates from the expected development, complex social processes have to take place in which the significance of the deviation is negotiated. Possible reasons for the deviation must be discussed and evaluated in terms of their plausibility. Based on the results of this discussion, a decision must be made on the kind of measures to be taken – of which retrieval would be the most extreme. If it has been decided that retrieval is necessary to ensure the safety and security of the site, information on the state of the repository is needed to safely plan retrieval activities.

Monitoring in this context means the temporary technical collection of data on the development of the repository and its social and technical surroundings in combination with the social processes of data interpretation (based on Hocke et al. 2012). To enable decision making, it must be decided which parameters need to be monitored and how this should be realized technically. There is no experience yet with the durability of monitoring techniques and technologies over a period of several decades to centuries. Any monitoring equipment installed underground cannot be repaired or exchanged once the repository has been closed. The precautionary principle demands that the possible occurrence of a situation in which retrieval is considered necessary be taken seriously. This means that precautionary actions must be taken now. This includes, for example, the incorporation of monitoring into the repository design, research on monitoring technologies, as well as debates about contextual factors for retrieval. So far, the debate has been limited and critically received in parts of the technical and scientific disposal community.

Further, taking into consideration current decision-making structures and calls for public participation, it can be assumed that the site operator will not be able to perform the interpretation of the data alone. Safety authorities, supervisory bodies, as well as scientific actors and stakeholders will have considerable interest in participating in such a task. In addition to the technical issue, an institutional setting must be found in which such a variety of actors can cooperate and make responsible and informed decisions.

2 Background

German high-level nuclear waste mainly consists of spent fuel rods from nuclear power plants and vitrified waste. When the first power plants were built in the mid-20th century, the waste was not considered a major problem. Rather, visions of a closed fuel cycle in which the waste would be reprocessed and reused dominated the debate. The fact that these reprocessing technologies also produce nuclear waste that requires an underground repository was not of particular interest. Since then, rather than being considered a minor problem, waste has come to dominate much of the debate on nuclear issues and has become a major subject of regulation. In several countries, entire public institutions are devoted to identifying a suitable repository site. This change can be interpreted as being due to reflexive modernization in which constant reinterpretations from resource to waste and vice versa take place and in which society attaches meaning to these wastes and the associated problems through processes of reflection (Kuppler 2017, based on Beck 1996 and Keller 2000).

As a result of these processes, attempts at solving the problem of "high-level waste disposal" have been strongly influenced by the debates on the use of nuclear power for energy production as well as by the political conflicts that have shaped this field in many countries ever since the first nuclear power plant was built. At the same time, this problem requires the attention of generations of researchers, policy makers, industry representatives, and citizens alike. The time span over which a repository is supposed to keep the waste safe and secure in Germany is one million years. This is of course not the time span over which some kind of human control is expected or deemed necessary. Still, also the time spans during which active or passive control must be realized are very long compared to planning periods for other industrial facilities. Visions of a quick succession of planning, construction, and closure of a so-called "maintenance-free" repository have proven unrealistic (Hocke et al. 2012).

The German Site Selection Act that came into effect in 2013 states that retrievability should be possible during the operation phase, i.e., during the period when the waste is brought underground. Further, the waste should be retrievable for 500 years after closure using mining techniques. In the meantime, more detailed requirements regarding the safety of a repository have been issued in Germany.[1] Despite careful planning, it can never be known whether a repository is really safe (cf. Berkhout 1991). In order to be able to retrieve the waste, the final storage containers in which the waste is stored would need to remain intact over this period. In order to retrieve the waste, information on the state of the underground repository and its surroundings is required. This information is usually gathered by technical monitoring. For an underground repository, monitoring techniques must be used that require no maintenance after closure of the repository. Such techniques are not readily available.

So far, few attempts have been made at defining the kinds of tasks that will need to be carried out in and at a repository over periods of 500 years or more and who will be responsible for performing and coordinating them. An example of such an attempt is the long-term stewardship program developed by the United States Department of Energy (e.g., U.S. Department of Energy 1999). This defines tasks that need to be accomplished over the next 100 years. Czada (2016) argues that long-term planning is a question of institutional settings. The scientific debate on who should decide in the future and what kind of information and competences future decision makers need in order to be able to act has just started (Kuppler/Hocke 2018). This anthology takes up the debate and intends to contribute from the perspective of various disciplines by approaching questions such as what the technical requirements for monitoring activities are, how legal studies interpret the institutional aspects, or what role public participation could play in long-term governance.

3 On the challenge of planning over long periods of time

3.1 Unbounded technologies

Technologies often have both intended and unintended effects that are unbounded in their spatial and temporal potency. Current political systems often struggle to find appropriate answers to problems associated with unintended effects. With regard to the hole in the ozone layer, coordinated action by many nations has been successfully implemented. Combating climate change, on the other hand, requires such profound changes in our economic activities that coordinated action is much more difficult to achieve. It also requires changes that take at least several decades to implement, such as transforming the energy

---

1 The Repository Safety Requirements Ordinance (Endlagersicherheitsanforderungsverordnung – EndlSiAnfV) came into force in October 2020. It specifies that retrieval should be possible without major effort until the start of closure of the repository (§ 13). Retrieval should be possible for 500 years after closure (§ 14).

system from one based on fossil fuels to one based on renewable energy. For example, a study published by the German Environment Agency (Umweltbundesamt – UBA) has suggested that the transformation of the German energy system from fossil-based to renewable energy should be completed by 2050 in order to reach the emission reductions aimed for (Klaus et al. 2010). Changes in government coalitions and indecisiveness in implementation can lead to several changes in the plan, which may to some extent question the original goal set. There seems to be a contradiction between the political focus on election periods and the need for long-term planning (Czada 2016). On the other hand, Gärditz (2013) argues that long-term planning is undemocratic since any democratic decision must be reversible. He argues that long-term planning involves balancing very different values: On the one hand, long-term interests that relate to the abstract bad that could happen in the future due to climate change; on the other hand, immediate threats such as, e.g., to biodiversity from the construction of renewable energy facilities. In his opinion, public institutions are not equipped to make such decisions.

Full reversibility is hardly ever a given with regard to technological infrastructures. Once the decision has been made to decommission a power plant and dismantling has started, it cannot simply be reversed. Calling for reversible decisions would imply maintaining the status quo. What reversibility can mean in the context of technological system changes and long-term tasks such as nuclear waste management is allowing for alternatives in a long-term plan and determining what resources and skills need to be developed so that institutions will be able to handle the technology and the associated societal impacts also in the future.

In order to be able to regulate the impact of technologies on society and the environment, information about their state is gathered with the help of technical monitoring. Mostly, environmental monitoring is about controlling the success of cleaner technologies on the one hand and controlling adherence to environmental regulations on the other. In many industrialized countries, it is standard procedure to equip industrial installations potentially emitting pollutants into the air or water with some kind of monitoring. Yet, for a planned nuclear waste repository, the debate on what kind of technical monitoring is needed and possible has only just begun. Monitoring in this context is discussed as one approach to dealing with the temporal unboundedness of the problem. Particularly technical experts often understand monitoring in this context as a technical tool to take care of all challenges and risks related to the underground infrastructure. In Germany and other countries planning for an underground repository, the preferred type was a maintenance-free underground repository[2] for a long time. For such a facility, technical monitoring would only be possible above ground, e.g. of air and water quality. In several countries, the debate has moved on and nowadays a type of repository with retrievability is planned for.[3]

To realize such a technical monitoring, two prerequisites need to be fulfilled: first, it must be decided which parameters are to be monitored and, second, technical solutions must be developed to actually monitor these parameters. Both prerequisites are not easy to fulfil. The parameters selected should provide the information needed to be able to judge whether everything develops as planned and, if it is not going as planned, whether this has implications for the safety of the repository that, in the worst case, would require retrieval of the waste. In what cases retrieval would be necessary is by no means clear. Thus, it is also not clear what would be suitable termination criteria. Existing monitoring techniques have never been used over such long periods of time without the possibility of maintenance, including replacement of parts. Developing such monitoring techniques that can reliably provide secure data over a very long time span without maintenance is a considerable engineering challenge.

An additional challenge is that some kind of long-term institution is needed that can interpret the data delivered by the monitoring technology and make decisions based on it. Interpreting the data and acting upon it means that the institutions must be able to decide whether a deviation from standard values implies a measurement error, an acceptable deviation, or whether measures need to be taken, such as retrieval of the waste. If action is considered necessary, the long-term institution needs to be able to activate other political and administrative institutions and civil society in its surroundings. Those institutions do not need to have detailed knowledge and skills to interpret monitoring data but need to cooperate in implementing measures deemed necessary by the long-term institution. To do so, the long-term institution must be able to muster resources even when public and political interest wanes. In addition to its technical abilities, the long-term institution must be able to keep up a dialogue with the public, in particular the local public (Kuppler/Hocke 2018). To fulfill these tasks, "learning institutions" are needed.[4] The currently responsible institutions

2 The term "maintenance-free" repository stands for the idea that with closure of the repository passive safety will be achieved, which eliminates the need for human control (e.g., Hocke et al. 2012; Kirch et al. 1990).

3 An underground repository with retrievability is constructed such that the waste can be retrieved with reasonable efforts if deemed necessary (e.g., Kommission Lagerung hoch radioaktiver Abfallstoffe 2016).

4 In Germany, the Repository Site Selection Act (Standortauswahlgesetz – StandAG) prescribes a "learning procedure" for the institutions involved. The StandAG itself has already proven to be

in Germany do not seem prepared to undertake such tasks and to cooperate in the necessary way. This shows particularly in the case of the Asse mine (see Hocke et al. 2016). What steps institutions must take to be prepared to take responsible decisions and actions in the future is an open question.

3.2 Current approaches

Currently, only one management concept dealing with long-term issues can be identified in the literature on nuclear waste management: The term "long-term stewardship" stands for a program developed by the U.S. Department of Energy (DOE) with the aim of coordinating the long-term management of nuclear sites, both military and others (U.S. Department of Energy 1999). It mainly lists tasks that need to be fulfilled if institutional control is to remain active. It is assumed that such institutional control can be provided for 100 years from now (U.S. Government Printing Office 2006). Experiences with the concept point to an important lesson: it is very difficult to foresee the economic costs of performing necessary tasks over a longer period of time and thus to allocate resources accordingly. At the same time, the concept does not address the important aspect of institutional requirements necessary to ensure that the tasks identified can actually be carried out (Kuppler/Hocke 2018).

This was also a topic of a keynote Metlay presented at an ITAS workshop in 2016, upon which this compendium is based (see section 4 in this introduction). Metlay pointed out that it is often assumed that such tasks are "self-implementing," i.e., that implementation is an inherent part of task identification (Metlay 2016). He argued that such an assumption is particularly problematic when it comes to monitoring as organizational issues play a major role, such as how to evaluate the information obtained from monitoring and how to decide whether to change plans when there is no agreement that a change is necessary. In his view, this would lead to two possibly contradictory characteristics that a monitoring organization would have to fulfill: it would need to be independent to be able to provide reliable and valid data and at the same time interdependent to organize discussion about the meaning of the data. Some kind of hybrid institution would be needed (Kuppler/Hocke 2018). Regarding the technical aspects, the key question in his view is: What should be measured and how can reliable measurements be taken?

---

a "learning system," as it has been adapted several times after consultations with experts and the public (Smeddinck 2017; see also Mbah/Brohmann 2021).

The questions raised in the opening talk can be understood as guiding questions for the whole workshop and this anthology.[5] First, the organizational question of the embeddedness of a monitoring institution and, second, the technical question of what and how. Czada (2016) also argues that long-term planning is a question of coordination and interaction, and thus of institutional settings. It can therefore be understood as a governance problem in which coordination and cooperation need to be organized (cf. Grande 2012; Mayntz 2009). In this anthology, Mbah adds to this debate by looking at public participation and Smeddinck et al. by looking at legal questions of long-term institutions.

The technical debate on the what and how has so far mainly been addressed in two projects funded by the European Union: MODERN and MODERN2020 (see the final conference report MODERN2020 2019). In their contribution to this anthology, Jobmann and Liebenstund present some results from the latter, with a particular focus on the interplay between the technical and the social. Still, from a technical point of view, many questions remain open, not least because appropriate monitoring also depends on the disposal concepts favored, which differ considerably between countries.

The solution to the technical problem is highly complex as the safety requirements for a monitoring approach do not necessarily coincide with the safety requirements for the repository structure. Thus, it is essential that a broad range of technical disciplines engage in an intensive discourse on the effects of the boundary conditions defined by each discipline on the feasibility of different monitoring approaches. On the one hand, this discussion should provide an overview of what is technically feasible and, on the other hand, the basis for what is socially feasible. However, there are no comprehensive overall approaches, neither technical nor social, to overcome the challenges associated with monitoring a repository underground. One example of such a challenge is the wireless transmission of measurement data through the repository rock, which is technically unresolved but a necessary prerequisite for achieving hermetic containment of the waste materials.

The identified need for institutions capable of collecting meaningful and reliable data and organizing a debate on the meaning of this data points to the need for interdisciplinary research. This type of research could help not only to understand the technical difficulties or the problems of institution building but also to analyze and advance knowledge about the socio-technical interplay in

5 Some of the contributions in this anthology are based on presentations made at the workshop. Others were added subsequently to broaden the perspectives on monitoring and long-term governance.

which such long-term governance settings develop.[6] This is of particular importance because the questions of what should be monitored and how monitoring should be organized cannot be answered by researchers alone (Bergmans et al. 2012).

4 From the idea to the anthology

This book is mainly based on research conducted within the ENTRIA project ("Disposal options for radioactive residues: Interdisciplinary analyses and development of evaluation principles").[7] In ENTRIA, interdisciplinary research was carried out on several topics related to nuclear waste management and mainly organized bottom up. This means that the research topics were developed in discussions during the project. The aim of most interdisciplinary research endeavors in ENTRIA was to investigate a particular problem by including different problem perspectives, rather than to develop interdisciplinary research methods (e.g., Brunnengräber et al. 2016; Röhlig et al. 2017; Köhnke et al. 2017).

Not long after the launch of ENTRIA, it became clear that a new political attempt would be undertaken to end the stalemate in decision-making on nuclear waste management in Germany. The StandAG was drafted and enacted, and the Commission on the Storage of High-Level Radioactive Materials was established. In its final recommendations, the Commission advocated for an underground repository with retrievability, without further elaborating on the details of what retrievability could mean in practice (Kommission Lagerung hoch radioaktiver Abfallstoffe 2016). Switzerland is one example of a country that has planned for retrievability. Since research in this field is still in its infancy, the ENTRIA members interested in this topic had to decide how to approach such a complex issue. An interdisciplinary and international workshop on "Technical Monitoring and Long-Term Governance" was considered a suitable approach. Such a workshop can provide a first insight into the state of the art and the potential contribution different disciplines can make to the academic discussion of this problem. It can help further debate on different aspects of the problem and provide a basis for discussion of new conceptual ideas.

6 An overview of approaches to socio-technical problems can be found, e.g., in Lösch 2012. For a debate on interdisciplinary research in nuclear waste management, see Smeddinck et al. 2016.

7 ENTRIA was a joint research project funded by the German Federal Ministry of Education and Research (BMBF, support code: 15S9082A) that ran from 2012 to 2017. For the program of the workshop and slides presented, see https://www.itas.kit.edu/veranstaltungen_2016_entria_temo.php [Accessed 14.01.2022].

The workshop was therefore attended by representatives from a variety of disciplines, including legal studies, science and technology studies (STS), geophysics/geology, radiochemistry, political science, institutional psychology, and engineering. They contributed ideas on technical possibilities for monitoring as well as research needs, perspectives on requirements for learning institutions, the role of uncertainty and ignorance, and experiences from the United States and Switzerland. Participants included: Dr. Anne Eckhardt (risicare), Prof. Dr. habil. Oliver Sträter (Kassel University), PD Dr. Stefan Böschen (ITAS at KIT), Dr. Anne Bergmans (University of Antwerp), Dr. Daniel Metlay (Member of the Senior Professional Staff of the U.S. Nuclear Waste Technical Review Board – NWTRB), Prof. Dr. Armin Grunwald (ITAS at KIT, former member of the German Commission on the Storage of High-Level Radioactive Materials, current member of the German National Citizens' Oversight Committee [Nationales Begleitgremium]), Prof. Dr. Horst Geckeis (INE at KIT).

All participants were invited to contribute to the present anthology, either in German or English. Furthermore, selected authors who could not attend the workshop were invited to contribute. A majority of the contributions were discussed at a meeting between the authors and editors. Each contribution was assigned to a participant who read the respective manuscript carefully and served as discussant during the meeting. This ensured that the contributions were written in a way that is comprehensible for an interdisciplinary audience. After the meeting, the contributions were revised and submitted to a final review by the editors. Contributions that could not be discussed at the meeting were subjected to an internal review. Thus, ideas from the workshop were scrutinized and further developed before being published in this compendium.

5 Overview of the contributions

The successful implementation of radioactive waste disposal depends on technical aspects, such as a sound strategy for ensuring safety and security, as well as scientific and engineering competences. Social aspects such as acceptance and trust of the different interest groups are another prerequisite. Monitoring is considered key to fulfilling all of these aspects. In their contribution, Jobman and Liebenstund discuss results of the MODERN2020 research project. Based on the guidelines developed within the MODERN project, MODERN2020 aims to provide the basis for the development and implementation of an efficient monitoring system for a final repository. The aim is to understand what should be monitored as part of the safety case and to provide a method for using monitoring results to support decision-making processes. Furthermore, MODERN2020 specifically aims to effectively involve local interest groups (stakeholders) in R&D monitoring activities. In this context, the general question of how successful cooperation can be organized and at what point of time in the process the involvement of local interest groups is meaningful and effective is to be clarified in dialogue with stakeholders.

Anne Eckhardt shows in her contribution that Switzerland follows a very specific concept of deep geological disposal for the management of radioactive wastes. This concept combines society's need for controllability with the strengths of classic approaches to radioactive waste disposal. At the disposal site, a pilot repository will be installed in the vicinity of the main repository where the majority of waste will be stored. The pilot repository must be representative of the main repository, but at the same time spatially and hydraulically separated. After closure of the main repository, provisions will be made for long-term monitoring in the pilot repository. In general, and particularly with regard to the pilot repository, the monitoring configuration in a geological repository requires not only technical and scientific knowledge but also supervision by members of civil society and the social sciences and humanities. From an ethical point of view, it is not possible to weigh the pros and cons of monitoring with regard to safety. Thus, a political decision must be made whether the decrease in uncertainty through monitoring outweighs the additional calculable risks associated with it.

Karl-Heinz Lux et al. argue that the intended monitoring concept for the planned repository needs to be discussed already now during the site selection process. The reason is that the monitoring concept can influence the selection criteria, e.g., due to the required space. They describe the basic idea behind the monitoring concept developed in the course of the ENTRIA project, which provides for a second level above the repository, from which monitoring boreholes reach the emplaced waste. They argue that with this setup, monitoring can take place during and after emplacement, thus contributing to intergenerational equity. Lux et al. also present some first results from their thermal, geomechanical, and fluid dynamic modeling of the development of such a two-level repository.

Any decision on the potential retrieval of high-level radioactive waste from the repository requires a sound data base. This data must be collected in a monitoring program that must be applied during the entire operational phase of the repository. It is yet unclear how monitoring will be integrated into the repository design. Volker Mintzlaff et al. discuss a generic repository design to meet this requirement. The approach presented considers a two-level mine in which monitoring is possible via boreholes drilled from the upper level to the closed emplacement drifts in the lower level. Due to the long operational phase, the monitoring devices must be replaceable. They further discuss the setup of the monitoring program in different host rocks and the trade-offs between the active control possibility with monitoring and long-term safety of the repository after the operational phase.

The call for a possibility to intervene in case the repository does not develop as planned has gained considerable importance in recent years. Directly related to this is the possibility to retrieve the stored containers. In this context, Thomas Hassel and Ansgar Köhler discuss various aspects of monitoring the technical barrier. To ensure safe handling of the containers in the course of a possible retrieval operation, damage to the technical barrier that could lead to a loss of integrity must be identified early. Otherwise, safe intervention is not possible. This contribution takes a look at the boundary conditions of monitoring storage containers in a deep repository. Limitations due to the geological and geotechnical surroundings are depicted, such as the lack of accessibility and maintainability of the sensor systems. The authors present possible solutions that allow for technical monitoring under the given circumstances.

The German Commission on the Storage of High-Level Radioactive Materials (Kommission Lagerung hoch radioaktiver Abfallstoffe, 2014–2016) was tasked with preparing a transparent site selection procedure and issuing recommendations for the revision of the German Repository Site Selection Act (StandAG). In his contribution, Armin Grunwald introduces the Commission's work on monitoring by presenting the ideas of process and repository monitoring as described in the Commission's final report. While no official process monitoring has been established by the responsible public authorities, he argues that one of the core tasks of the National Citizens' Oversight Committee (Nationales Begleitgremium – NBG) is to monitor and critically assess the ongoing site selection process.

Melanie Mbah discusses the failure of Germany's representative democracy to perform interest aggregation regarding nuclear waste management in today's pluralistic and fragmented society. Dissatisfied citizens often challenge top-down decision-making in increasingly complex problem structures. Democracies in general seem less able to make robust decisions that gain acceptability and legitimacy. To remedy this, the use of participatory elements in decision-making processes is continuously expanding in order to represent different views and interests and to build legitimacy. This raises new challenges regarding participation and democratic standards and principles. Mbah discusses how different actors consider participation necessary and that their perspectives on what participation means and why and how it should be implemented need to be taken into account. The institutionalization of participatory elements might be a solution, which also means new ways of working for institutions and authorities. Nuclear waste governance in Germany is a good example of current efforts in this direction.

The task of realizing a safe repository for highly radioactive wastes – colloquially also known as atomic waste – for one million years is unprecedented in Germany, as Ulrich Smeddinck and Franziska Semper state in their contribution. Also unprecedented is the challenge for legal and administrative bodies to supervise a final repository for as long as possible. This means acknowledging the weakness and transience of human civilization while at the same time considering a possible capacity for adaptation and renewal. How can such long-term governance be realized from the perspective of legal studies? This contribution clarifies the legal understanding of the term "governance," outlines existing legislative approaches that offer a long-term perspective, and develops first ideas on organizations and procedures that take into account future generations with regard to the long-term institutional supervision of a repository. Scenarios of long-term governance are developed, first under stable conditions and second in a dissolving state. Finally, future research perspectives are presented.

For several decades, increasing attention has been paid to the problem of non-knowledge. This is why the precautionary principle has become a principal framework for EU regulatory efforts. Stefan Böschen argues in his article that these questions become more relevant in situations where ambitious socio-technological solutions are to be created. These solutions evolve as they pose a "linked series of socio-technical problems," as Paul Edwards puts it to describe their co-evolutionary development, which must balance physical, technical, organizational, legislative, and societal factors. Therefore, the complex processes of problem articulation and solving must be taken into account. He puts forward the thesis that these processes can be described as the formation of knowledge-regimes. In such regimes, problems are constructed by providing problem descriptions with related indicators, which are then selected. Therefore, indicator policy is the way to influence the problem-solving process.

The safety of disposal facilities for radioactive wastes strongly depends on human influences, since even a long-term disposal strategy is ultimately a human decision and interpretation. Oliver Sträter discusses the high demands this places on the operator and its surroundings. This is particularly true in the design and planning phase, as decisions once taken are rarely revised despite better knowledge. This phenomenon is known as "drift into failure" and requires consideration of human aspects in decision-making and intervention processes. This contribution describes the necessary aspects of human and organizational action for long-term disposal that allow timely responses to possible safety problems. It goes into the phenomenology of human errors and describes the processes needed in error management to strengthen the resilience of a disposal solution. Here, Sträter addresses the requirements on the system as a whole and not only for the operator.

References

Beck, U. (1996): Das Zeitalter der Nebenfolgen und die Politisierung der Moderne. In: Beck, U.; Giddens, A.; Lash, S. (eds.): Reflexive Modernisierung. Eine Kontroverse. 5th edition. Frankfurt am Main, pp. 19–112

Bergmans, A.; Elam, M.; Simmons, P.; Sundqvist, G. (2012): Perspectives on Radioactive Waste Repository Monitoring. TATuP 21(3), pp. 22–28

Berkhout, F. (1991): Radioactive Waste. Politics and Technology. London/New York

Brunnengräber, A.; Hocke, P.; Kalmbach, K.; König, C.; Kuppler, S.; Röhlig, K.-J.; Smeddinck, U.; Walther, C. (2016): Grenzwerte beim Umgang mit radioaktiven Reststoffen. ITAS-ENTRIA-Arbeitsbericht 2015–01, Karlsruhe

Czada, R. (2016): Planen und Entscheiden als Steuerungsaufgabe und Interaktionsproblem. In: Kamp, G. (ed.): Langfristiges Planen: Zur Bedeutung sozialer und kognitiver Ressourcen für nachhaltiges Handeln. Berlin, pp. 215–249

Gärditz, K.F. (2013): Zeitprobleme des Umweltrechts – Zugleich ein Beitrag zu interdisziplinären Verständigungschancen zwischen Naturwissenschaften und Recht. Europäisches Umwelt- und Planungsrecht 11(1), pp. 2–16

Grande, E. (2012): Governance-Forschung in der Governance-Falle? Eine kritische Bestandsaufnahme. Politische Vierteljahresschrift 53(4), pp. 565–592

Hocke, P.; Bechthold, E.; Kuppler, S. (2016): Rückholung der Nuklearabfälle aus dem früheren Forschungsbergwerk Asse II. Dokumentation einer Vortragsreihe am Institut für Technikfolgenabschätzung und Systemanalyse (ITAS) (mit Beiträgen von Detlev Möller, Beate Kallenbach-Herbert, Silvia Stumpf, Volker Metz). KIT Scientific Working Papers Nr. 47, Karlsruhe

Hocke, P.; Bergmans, A.; Kuppler, S. (2012): Guaranteeing Transparency in Nuclear Waste Management: Monitoring as Social Innovation. TATuP 21(3), pp. 10–14

Keller, R. (2000): Der Müll in der Öffentlichkeit. Reflexive Modernisierung als kulturelle Transformation. Ein deutsch-französischer Vergleich. Soziale Welt 51(3), pp. 245–266

Kirch, N.; Brinkmann, H.U.; Brücher, P.H. (1990): Storage and final disposal of spent HTR fuel in the Federal Republic of Germany. Nuclear Engineering and Design 121(2), pp. 241–248

Klaus, T.; Vollmer, C.; Werner, K.; Lehmann, H.; Müschen, K. (2010): Energy target 2050: 100 % renewable electricity supply. Dessau-Roßlau

Köhnke, D.; Reichardt, M.; Semper, F. (2017): Zwischenlagerung hoch radioaktiver Abfälle. Wiesbaden

Kommission Lagerung hoch radioaktiver Abfallstoffe (2016): Verantwortung für die Zukunft. Ein faires und transparentes Verfahren für die Auswahl eines nationalen Endlagerstandortes. K-Drs. 268

Kuppler, S. (2017): Effekte deliberativer Ereignisse in der Endlagerpolitik. Wiesbaden

Kuppler, S.; Hocke, P. (2018): The role of long-term planning in nuclear waste governance. Journal of Risk Research. Published online, 18 Apr 2018, pp. 1343–1356

Lösch, A. (2012): Techniksoziologie. In: Maasen, S.; Kaiser, M.; Reinhart, M.; Sutter, B. (eds.): Handbuch Wissenschaftssoziologie. Wiesbaden, pp. 251–264

Mayntz, R. (2009): Von politischer Steuerung zu Governance? In: Mayntz, R. (ed.): Über Governance. Institutionen und Prozesse politischer Regelung. Frankfurt am Main, pp. 105–120

Mbah, M.; Brohmann, B. (2021): Das Lernen in Organisationen. Voraussetzung für Transformationsprozesse und Langzeit-Verfahren. In: Brohmann, B.; Brunnengräber, A.; Hocke, P.; Isidoro Losada, A.M. (eds.): Robuste Langzeit-Governance bei der Endlagersuche. Soziotechnische Herausforderungen im Umgang mit hochradioaktiven Abfällen. Bielefeld, pp. 387–412

Metlay, D. (2016): Organizations Matter: Monitoring and Long-Term Governance. Presentation at the Workshop "Technical Monitoring and Long-term Governance," Karlsruhe, 18.–19.10.2016; http://www.itas.kit.edu/downloads/veranstaltung_2016_entria_temo_metlay.pdf [Accessed 10.01.2022]

MODERN2020 (2019): Final Conference Proceedings. Deliverable 6.3, Horizon2020-Project MODERN2020 (ed.), Paris, April 2019, pp. 280–283; http://www.Modern2020-_D6.3_PU_Conference_proceedings_FINAL-web.pdf [Accessed 31.01.2022]

Röhlig, K.-J.; Häfner, D.; Lux, K.-H.; Hassel, T.; Stahlmann, J. (2017): Einschluss oder Zugriff. GAIA 2, pp. 114–117

Smeddinck, U. (2017): Die Fortentwicklung des StandAG – Novellierungen, Beispiele, Reflektionen. Europäisches Umwelt- und Planungsrecht 15(3), pp. 195–205

Smeddinck, U.; Kuppler, S.; Chaudry, S. (eds.) (2016): Inter- und Transdisziplinarität bei der Entsorgung radioaktiver Reststoffe: Grundlagen – Beispiele – Wissenssynthese. Wiesbaden

U.S. Department of Energy (1999): From Cleanup to Stewardship – A Companion Report to Accelerating Cleanup: Paths to Closure. Washington, DC

U.S. Government Printing Office (2006): Code of Federal Regulations. Title 40 – Protection of the Environment. 40 CFR 191.14a, Washington, DC

Michael Jobmann, Anna-Laura Liebenstund

Monitoring im europäischen Kontext. Die Projekte MoDeRn und MODERN2020

1 Einleitung

Die erfolgreiche Realisierung eines Endlagerprogramms für hochradioaktive und langlebige Abfälle beruht sowohl auf technischen Aspekten wie einer fundierten Sicherheitsstrategie und wissenschaftlicher und ingenieurtechnischer Kompetenz als auch auf sozialen Aspekten wie Akzeptanz und Vertrauen vonseiten der Interessengruppen, z. B. lokal ansässiger Bürger. Der Aspekt der Endlagerüberwachung zumindest während der Betriebsphase und vielleicht auch nach dem Verschluss des Endlagers durch sogenanntes „Monitoring" gilt in Fachkreisen wie der IAEA oder der NEA als Schlüssel, wenn es darum geht, beiden Aspekten gerecht zu werden.

Entsprechende Monitoring-Technologien werden bereits seit mehreren Jahrzehnten entwickelt und momentan vor allem zur Bewertung der Betriebssicherheit von Endlagerbergwerken eingesetzt. In einigen Fällen wurden Aspekte eines Monitorings auch in nationale Regularien übernommen. Beispielsweise beinhaltet der US Code of Federal Regulations (CFR) (US NRC 2012) spezifische Anforderungen zur Prüfung des Endlagerverhaltens (Repository Performance Confirmation).

Spätestens seit der Jahrtausendwende sind Aspekte eines Monitorings auch in Dokumenten der IAEA verankert. In der IAEA-TECDOC-1208 (IAEA 2001) werden beispielsweise Anwendungszwecke eines Monitorings in einem Endlager diskutiert, insbesondere im Hinblick auf die verschiedenen Zeitabschnitte eines Endlagers, beginnend mit der Standorterkundung bis hin zu einem möglichen Monitoring nach Verschluss des Endlagers. Es wird darin auch diskutiert, wie die Informationen, die ein Monitoring liefert, verwendet werden könnten und welche Techniken für ein Monitoring-System zum Einsatz kommen könnten.

Auf der Grundlage dieser Diskussionen wurde dann auf europäischer Ebene ein erstes Forschungsprojekt gestartet, das sogenannte „European Thematic Network on Monitoring" (ETN) (EC 2004). Dieses Projekt wurde von elf Partnern aus zehn europäischen Ländern durchgeführt, mit dem Ziel, die Rolle eines Monitorings im Zuge der schrittweisen Entwicklung eines Endlagers für hochradioaktive Abfälle zu erörtern. Dabei wurde deutlich, dass es sinnvoll ist, das Thema Monitoring auf europäischer Ebene weiterzuentwickeln, um insbesondere ein gemeinsames international anerkanntes Verständnis für die Entwicklung eines Monitoring-Programms für ein Endlager zu schaffen. Diesem Gedanken folgend gelang es, ein Projekt bei der Europäischen Kommission zu lancieren, das MoDeRn-Projekt, das darauf abzielte, erstmalig so etwas wie eine Rahmenrichtlinie für ein Monitoring-Programm auf internationaler Ebene zu erstellen. Ziel war es, dieses gemeinsame Verständnis zu dokumentieren, relevante Monitoring-Technologien zu identifizieren, den Stand ihrer Technik zu eruieren und Sichtweisen von Stakeholdern zu dokumentieren, um eine Grundlage für die Entwicklung individueller Monitoring-Programme zu schaffen (MoDeRn 2013a).

2 Das Projekt MoDeRn

Das Projekt MoDeRn (*Monitoring Developments for Safe Repository Operation and Staged Closure*; www.modernfp7.eu) war ein Verbundprojekt mit einer Laufzeit von vier Jahren (2009–2013) und Teil des 7. Rahmenprogramms der Europäischen Atomgemeinschaft (EURATOM). Im Rahmen dieses Projekts wurde von 18 Organisationen (mehrheitlich Vertreter der jeweiligen nationalen Betreibergesellschaften, aber auch Sozialwissenschaftler und Vertreter gesetzgebender Institutionen) aus zwölf verschiedenen Ländern eine erste Rahmenrichtlinie zur Entwicklung und Umsetzung verschiedener Monitoring-Aktivitäten für Endlager für hochradioaktive Abfälle und ausgediente Brennelemente (HAW/SF) auf europäischer Ebene erarbeitet. Die Rahmenrichtlinie stützt sich auf Erfahrungen, die in den Endlagerprogrammen der Länder gemacht wurden, und entwickelt erste Ansätze zur Einbindung der interessierten Öffentlichkeit.

Im Zuge des Projekts wurde ein struktureller Ansatz, der sogenannte „MoDeRn Monitoring Workflow“, zur Entwicklung und Implementierung eines Monitoring-Programms für ein Endlager erstellt (MoDeRn 2013b). Dieser Workflow beschreibt den schrittweisen Prozess beginnend mit der Identifizierung der Anforderungen an ein Monitoring über die Umsetzung dieser Anforderungen in ein spezifiziertes Monitoring-Programm bis zur Analyse und Bewertung der Ergebnisse mit Blick auf die Sicherheitsfunktionen der geologischen und geotechnischen Barrieren.

Dementsprechend werden drei Schlüsselbereiche unterschieden:

- *Zielsetzungen und Parameter*: Festlegung von Zielen eines Monitoring-Programms und Identifizierung von Prozessen, deren Monitoring dazu dient, diese Zielsetzungen zu erreichen. Bestimmung von Parametern, deren Beobachtung es erlaubt, die relevanten Prozesse zu charakterisieren.

- *Monitoring-Programm und Ausgestaltung*: Analyse der Anforderungen an Monitoring-Systeme, Analyse der verfügbaren Technologien, um diesen Anforderungen gerecht zu werden, und Identifizierung der Parameter, die messtechnisch in geeigneter Form erfasst werden können.
- *Implementierung und Überwachung*: Durchführung des Monitoring-Programms und Auswertung der Ergebnisse als Grundlage für Entscheidungsfindungen.

Um dieses Verfahren zu testen, wurden anhand des Workflows Monitoring-Konzepte für drei verschiedene Endlagerkonzepte, die für drei verschiedene Wirtsgesteine entwickelt wurden, erarbeitet: das französische Konzept für ein Endlager in Ton, das finnische Konzept für ein Endlager in Granit und das damalige deutsche Konzept für ein Endlager im Salz (MoDeRn 2013c).

Neben diesen konzeptionellen Arbeiten wurden auch technologische Entwicklungen durchgeführt. Dies betraf die Sensortechnologie, z. B. im Bereich der faseroptischen Sensorik und der seismischen Tomographie, wie auch Techniken zur Datenübertragung. Ein Schwerpunkt der Entwicklung lag dabei im Bereich der kabellosen Datenübertragung innerhalb eines Endlagers und aus einem Endlager heraus. Hintergrund ist, dass Kabelführungen immer eine Schwächezone darstellen und es schwer nachweisbar ist, dass beispielsweise Kabelführungen durch eine Barriere nicht als bevorzugte Fließpfade für Flüssigkeiten dienen und so die Langzeitsicherheit des Endlagers nachhaltig gefährdet werden könnte. Aus diesem Grund wurden beispielsweise für das deutsche Konzept drei wesentliche Prinzipien aufgestellt:

(1) keine Kabelführungen durch Abdichtbauwerke
(2) keine Kabelführungen durch versetzte Strecken
(3) keine Kabelführungen aus dem Endlager zur Erdoberfläche

Diese Prinzipien machen die Entwicklung von Systemen mit kabellosem Datentransfer unabdingbar.

Im Rahmen des MoDeRn-Projekts wurden zwei unterschiedliche Systeme für die kabellose Datenübertragung entwickelt. Ein System für kurze Übertragungsdistanzen unter Verwendung hochfrequenter elektromagnetischer Wellen und ein System für lange Übertragungsdistanzen mit niederfrequenten Wellen (MoDeRn 2013d). Das System für kurze Reichweiten wurde federführend von der Firma Aitemin in Spanien entwickelt. Dieses System besteht aus einer Reihe von einzelnen Modulen, die, etwa handgroß, in der Lage sind, z. B. Feuchtigkeit, Porenwasserdruck und Totaldruck an einem Punkt zu messen und die Messergebnisse dann an eine Empfangseinheit zu übertragen. Die Stromversorgung läuft dabei über eine langlebige Batterie. Die Übertragungsdistanz beträgt einige Meter. Mit einem solchen System besteht die Möglichkeit, Messdaten durch ein Abdichtbauwerk hindurch oder aus einem Abdichtbauwerk heraus nach außen zu übertragen.

Das System für lange Übertragungsdistanzen wurden von dem niederländischen Projektpartner NRG entwickelt. Dieses System wurde in-situ im belgischen Untertagelabor HADES getestet. Dort ist es gelungen, Messdaten aus dem Untertagelabor durch die darüber liegenden Gesteinsformationen zur Erdoberfläche zu übertragen. Die Distanz beträgt 225 m. Grundsätzlich eröffnet sich damit die Möglichkeit, Messdaten auch aus einem bereits verschlossenen Endlager zur Erdoberfläche zu übertragen.

Die Ergebnisse des Projekts sind in einer Reihe von Berichten dokumentiert, die u. a. in der Mediathek auf der Internetseite der DBE TECHNOLOGY GmbH zum Download zur Verfügung stehen (www.dbe-technology.de).

3 Das Projekt MODERN2020

Aufbauend auf den Ergebnissen des MoDeRn-Projekts hat das von der Europäischen Kommission geförderte Projekt MODERN2020 (*Development and Demonstration of monitoring strategies and technologies for geological disposal*; www.modern2020.eu) das Ziel, die Grundlagen zur Entwicklung und Umsetzung eines effizienten Programms für ein Endlager-Monitoring bereitzustellen, das die spezifischen Anforderungen nationaler Endlagerprogramme berücksichtigt. Die Projektlaufzeit beträgt vier Jahre (2015–2019) und wird von 28 Partnern aus zwölf Ländern, koordiniert von der französischen ANDRA (Agence Nationale pour la Gestion des Déchets Radioactifs), durchgeführt. Die Arbeit ermöglicht es fortgeschrittenen Entsorgungsprogrammen, Monitoring-Systeme zu entwerfen, die nach Inbetriebnahme eines Endlagers eingesetzt werden können, und unterstützt weniger entwickelte Entsorgungsprogramme, wenn es darum geht, von Grund auf Monitoring-Systeme neu zu konzipieren. Ziel ist es zu verstehen, was ein Monitoring im Zusammenhang mit einem Langzeitsicherheitsnachweis für ein Endlager leisten kann, und eine Methodik zur Verfügung zu stellen, wie Ergebnisse des Monitorings verwendet werden können, um Prozesse der Entscheidungsfindung zu unterstützen. Weiterhin soll erarbeitet werden, welche Handlungsoptionen als Reaktion auf Monitoring-Ergebnisse möglich und sinnvoll sind.

Einer der Schwerpunkte von MODERN2020 liegt auf der Entwicklung von Konzepten zu einem Monitoring geotechnischer Barrierensysteme (EBS – Engineered Barrier System), ohne deren Dichtwirkung zu beeinträchtigen. Dies ist u. a. ein Punkt, der von dem deutschen Projektpartner bearbeitet wird. Im Rahmen eines Verbundprojekts, das die DBE TECHNOLOGY GmbH in Zusammenarbeit mit der BGR Hannover und der GRS in Braunschweig durchführt, wird die Methodik eines Sicherheitsnachweises für ein Endlager in einer deutschen Tonformation entwickelt (Jobmann et al. 2016). Für das darin erarbeitete Verschlusskonzept soll überlegt werden, in welcher Form ein Monitoring hilfreiche Informationen über das Barrierensystem liefern kann.

Innovative Monitoring-Techniken sollen im Zuge des Projekts entwickelt und verbessert werden, insbesondere auf dem Gebiet der kabellosen Datenübertragung und alternativer Stromversorgung, die die Abwärme des eingelagerten Abfalls als Energiequelle nutzt und so nicht auf traditionelle Batterien angewiesen ist. Praktische Tests im Rahmen von In-situ-Großversuchen sollen die Einsatzfähigkeiten demonstrieren.

Abbildung 1: Struktureller Aufbau des Projekts MODERN2020

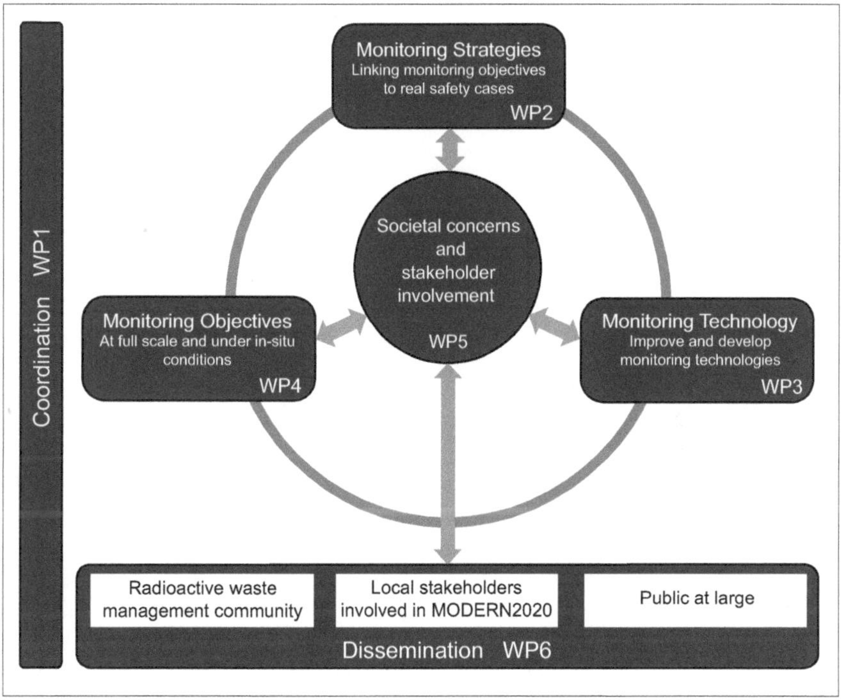

Coordination WP1
Monitoring Strategies
Linking monitoring objectives to real safety cases
WP2
Societal concerns and stakeholder involvement
WP5
Monitoring Objectives
At full scale and under in-situ conditions
WP4
Monitoring Technology
Improve and develop monitoring technologies
WP3
Radioactive waste management community
Local stakeholders involved in MODERN2020
Public at large
Dissemination WP6

Quelle: angelehnt an Bertrand 2017

Die Ergebnisse des Projekts sollen der interessierten Fachöffentlichkeit im Rahmen einer Abschlusskonferenz vorgestellt werden. Zusätzlich ist von der

MODERN2020-Gruppe auch ein einwöchiger Trainingskurs für Ingenieure, fortgeschrittene Doktoranden und promovierte Wissenschaftler geplant. Dieser Kurs soll auch den Besuch eines der europäischen Untertagelabors beinhalten.

Ein spezielles Ziel im Rahmen von MODERN2020 ist es, lokale Interessengruppen (Stakeholder) effektiv in die FuE-Monitoring-Aktivitäten einzubinden. In diesem Zusammenhang soll im Dialog mit Stakeholdern zum einen die generelle Frage geklärt werden, wie eine Dialog zwischen Bürgern und Betreibern aussehen könnte, und zum anderen, ab welchem Zeitpunkt eine Einbindung lokaler Interessengruppen sinnvoll und effektiv ist. Dieser Punkt nimmt auch eine zentrale Position innerhalb des Projekts ein, wie Abbildung 1 (siehe oben) zu entnehmen ist.

4 Projektübergreifende Untersuchungen zur Einbindung von Stakeholdern

4.1 Theoretischer Ansatz der Bürgerpartizipation in MoDeRn und MODERN2020

Das Ziel der Einbindung von lokalen Bürgern, die vom Bau eines Endlagers in der Umgebung betroffenen sein könnten, ist zu testen, inwiefern diese in die Forschungsarbeit eines komplexen technischen Projekts einbezogen werden können. Dieser Ansatz steht unter der generellen Annahme, dass Monitoring ein soziotechnisches Thema ist, welches durch eine Vielzahl von Faktoren beeinflusst wird. Aus dieser Sichtweise sind die unterirdische Entsorgung und das entsprechende Monitoring nicht nur eine Frage des technischen Know-hows. Als gesellschaftliche Einflussfaktoren wären zum Beispiel der politische, legislative und finanzielle Hintergrund zu nennen, bei den technischen Faktoren spielt neben dem technischen Know-how unter anderem auch der Stand der Technik eine entscheidende Rolle. Die Perspektive der Science and Technology Studies, von der im Projekt ausgegangen wird, verfolgt den Ansatz, dass Laienwissen qualitativ anders, aber weder falsch noch nutzlos ist (Bucchi/Neresini 2008). Aus dieser Perspektive heraus ist die Trennung zwischen technischen „Experten" und „Laien" fehlerhaft. Das Ziel in MODERN2020 ist nun zu vertiefen, was das „lokale Wissen" der Bürger ist (Wynne 1989), um herauszufinden, wie man ein „sozial robustes" Monitoring-System schaffen kann, in dem sich die Belange und Interessen der Gesellschaft widerspiegeln (Nowotny 2003). Wir versuchen „Upstream Engagement" in der Praxis, das heißt, die Einbindung der Bürger bereits in die Entwicklungsphase des Monitorings, bevor die Technologie vollständig entwickelt ist, da wir davon ausgehen, dass auf der Basis von Dialog gegenseitiges Vertrauen und auch ein Lerneffekt zwischen Bürgern und Experten erreicht werden kann (Wilsdon/Willis 2004). Langfristiges Ziel ist es, durch die Einbindung von Bürgern in beiden Projekten, MoDeRn und MODERN2020, Rückschlüsse zu ziehen, wie man Bürger in andere technisch komplexe EU-Forschungsprojekte einbinden kann.

4.2 Projekt MoDeRn: Was kann Monitoring überhaupt leisten?

Die Forschungsfrage bei der Analyse von Monitoring als soziotechnisches Problem im Vorläuferprojekt MoDeRn bestand darin, herauszufinden, was Monitoring leisten kann und wie verschiedene Stakeholder – technische Experten aus dem Bereich der Entsorgungsdienstleister und lokale, potenziell betroffene Bürger – Monitoring einschätzen. Um dies herauszufinden, wurden zahlreiche Interviews und Fokusgruppendiskussionen geführt. Ergebnis dieser Untersuchungen war, dass Experten und Bürger generell eine unterschiedliche Sicht auf Monitoring haben. Für die Experten besteht der Zweck des Monitorings in „Performance Confirmation", also der Bestätigung, dass die Prozesse innerhalb des Endlagers wie erwartet verlaufen. Laien hingegen sehen Monitoring viel mehr als eine Vorsichtsmaßnahme: Für sie geht es darum zu überprüfen, wie es um die Sicherheit des Endlagers steht. Dies zeigt, dass Bürger ein „demütiges" Verhalten von Experten gegenüber einem Endlagerprojekt erwarten. Sie erwarten, dass gerade die Verantwortlichen für die Endlagerentwicklung offen für die Infragestellung ihrer Arbeit bleiben – und Monitoring könnte ein System sein, das diese kritische Selbstanalyse in Bezug auf das entwickelte Endlagersystem in gewisser Weise ermöglicht, da es Echtzeitdaten aus dem inneren des Endlagers bereitstellt und so diesen „Realitäts-Check" liefern kann.

4.3 Projekt MODERN2020: Wie kann die interessierte Öffentlichkeit in die Entwicklung eines Monitoring-Programms eingebunden werden?

Aufbauend auf diesem Ergebnis hat die Beteiligung von Bürgen in MODERN2020 das Ziel, lokale Bürger aus mehreren EU-Ländern in die Entwicklung eines Endlager-Monitoring-Programms einzubinden, unter Berücksichtigung der verschiedenen soziotechnischen nationalen Kontexte. Auf Projektebene werden Repräsentanten der lokalen Gemeinden in projektinterne Workshops und Meetings eingebunden; gleichzeitig finden auf lokaler Ebene Diskussionsgruppen statt. Die beiden ersten dieser Diskussionsgruppen wurden im Mai 2016 in Belgien und Schweden durchgeführt. In Belgien fand der Workshop in der Gemeinde Mol/Dessel statt. Hier befindet sich seit den 1970er Jahren das Untertagelabor Hades. Zudem wird hier in den kommenden Jahren ein übertägiges Lager für leichtradioaktiven Abfall gebaut. Da sich die Gemeinden deshalb als mögliche Kandidaten für eine Endlagerstätte sehen, sind sie daran interessiert, Forschung bezüglich der Lagerung von hochradioaktiven Abfällen zu verfolgen. Die schwedische Gemeinde Osthammer, so bereits beschlossen, wird in ihrer nahen Umgebung ein Endlager aufnehmen. Zudem hat kürzlich die finnische Gemeinde Eurajoki ihr Einverständnis gegeben, am Projekt MODERN2020 teilzunehmen.

Die Einbindung der Bürger in MODERN2020 stand anfangs unter den Leitfragen „Was verstehen Bürger unter erfolgreicher Partizipation“ und „Was wissen Bürger bereits über das Thema Monitoring, was wollen sie noch erfahren?“ Bei der Auswertung dieser Fragen wurde bisher deutlich, dass Bürger großes Interesse am Thema Monitoring haben und es gerne weiterverfolgen möchten. Da die einbezogenen Gemeinden sich bereits seit geraumer Zeit mit dem Thema Endlagerung radioaktiver Abfälle auseinandersetzen, sehen sie sich als eine Art Experten. Aufgrund dieses Wissen sehen sie sich dazu in der Lage, zwischen der breiten Öffentlichkeit und Fachleuten als eine Art Vermittler fungieren zu können. Was jedoch momentan noch ein Problem bereitet, ist die Abstraktheit des Begriffs „Monitoring“, was das Verständnis stark erschwert. Da sich das Thema Monitoring noch im FuE-Stadium befindet, gestaltet es sich für Laien, scheinbar im Gegensatz zur Theorie, die eine frühe Einbindung von Bürgen empfiehlt, als schwierig, kritische, technisch detaillierte Fragen zum Thema zu stellen, wozu sie aber gerne in der Lage wären – was Bürger in jedem Fall ablehnen, ist das bloße „Abstempeln“ von bereits getroffenen Entscheidungen. Aus diesem Kontext heraus stellt sich nun die Frage, wann der beste Zeitpunkt ist, um Bürger in ein Forschungsprojekt einzubinden. Um das Thema Monitoring zu konkretisieren, wird nun auf der Grundlage dieser Zwischenergebnisse gemeinsam mit den Bürgern ein Monitoring-Handbuch entwickelt, indem die nationalen Monitoring-Programme und -Strategien erläutert werden. Der Vorteil eines Handbuchs besteht darin, dass die unterschiedlichen Interessen der verschiedenen Parteien bezüglich des Themas Monitoring nebeneinander dargestellt werden können. So können die verschiedenen Facetten von Monitoring als Technologie dargestellt werden, auch wenn das Thema aus unterschiedlichen Perspektiven betrachtet wird. Ziel ist es, anschaulich darzulegen, wie und warum Monitoring-Programme entwickelt werden. Das im Handbuch bereitgestellte Detailwissen soll Bürger ermutigen, kritische, vielleicht auch technischere Fragen zu stellen.

5 Zusammenfassung

Das Thema Monitoring eines Endlagers für hochradioaktive Abfälle gewinnt zunehmend an Bedeutung. Wurden erste Ansätze zunächst von einzelnen Endlagerinstitutionen erarbeitet, so wird seit einigen Jahren dieses Thema auch auf europäischer Ebene im Rahmen von Verbundprojekten intensiv diskutiert. Das Projekt MoDeRn (2009–2013) war ein solches Verbundprojekt, in dessen Verlauf eine erste Rahmenrichtlinie zur Entwicklung und Umsetzung von Monitoring-Programmen für HAW-Endlager entwickelt wurde. Die Rahmenrichtlinie stützt sich auf Erfahrungen, die in den Endlagerprogrammen der Partnerländer gemacht wurden, und beschäftigte sich erstmals mit der Frage einer möglichen Einbindung der interessierten Öffentlichkeit. Aufbauend auf den Ergebnissen dieses Projekts zielt das sich daran anschließende und seit 2015 laufende Projekt MODERN2020 darauf ab, die Grundlagen zur Entwicklung und Umsetzung eines effizienten Programms für ein Endlager-Monitoring bereitzustellen, das die spezifischen Anforderungen nationaler Endlagerprogramme berücksichtigt. Ein übergeordnetes Ziel ist es, zu verstehen, was ein Monitoring im Zusammenhang mit einem Langzeitsicherheitsnachweis für ein Endlager leisten kann. In diesem Zusammenhang sollen die Möglichkeiten und Grenzen eines Monitorings im Dialog mit interessierten Bürgern erörtert werden, mit dem Ziel, eine von möglichst vielen Seiten getragene Entwicklung eines Monitoring-Programms für ein HAW-Endlager zu gestalten.

Danksagung

Das Projekt MoDeRn wurde durch das Siebte Rahmenprogramm der Europäischen Atomgemeinschaft (FP7/2007–2011) unter der Finanzhilfevereinbarung Nr. 232598 gefördert. Das Projekt MODERN2020 wurde aus dem Euratom-Forschungs- und Ausbildungsprogramm 2014–2018 unter der Finanzhilfevereinbarung Nr. 662177 finanziert. Die Forschungsarbeiten der DBE TECHNOLOGY GmbH wurden mit finanzieller Unterstützung des BMWi vertreten durch den Projektträger Karlsruhe (KIT) durchgeführt.

Literatur

Bertrand, J. (2017): Presentation and discussion of the EU supported project MODERN2020, IAEA Technical Meeting to Establish a Working Group on the Use of Monitoring Programmes in the Safe Development of Geological Disposal Facilities for Radioactive Waste, 18–21 December 2017, Wien

Bucchi, M.; Neresini, F. (2008): Science and Public Participation. In: Hackett, E.J.; Amsterdamska, O.; Lynch, M.; Wajcman, J. (Hg.): The Handbook of Science and Technology Studies. Cambridge, MA, S. 449–72

EC – European Commission (2004): Thematic Network on the Role of Monitoring in a Phased Approach to Geological Disposal of Radioactive Waste. European Commission Project Report EUR 21025 EN. Luxemburg

IAEA – International Atomic Energy Agency (2001): Monitoring of Geological Repositories for High Level Radioactive Waste. IAEA-TECDOC-1208, Wien

Jobmann, M.; Bebiolka, A.; Jahn, S.; Lommerzheim, A.; Maßmann, J.; Meleshyn, A.; Mrugalla, S.; Reinhold, K.; Rübel, A.; Stark, L.; Ziefle, G. (2016): Sicherheits- und Nachweismethodik für ein Endlager in einer Tongesteinsformation in Deutschland. Projekt ANSICHT: Methodik und Anwendungsbezug eines Sicherheits- und Nachweiskonzeptes für ein HAW-Endlager im Tonstein. Synthesebericht. Peine u. a. O.

MoDeRn (2013a): Monitoring During the Staged Implementation of Geological Disposal: The MoDeRn Project Synthesis. Deliverable D6.1

MoDeRn (2013b): MoDeRn Monitoring Reference Framework Report. Deliverable D1.2

MoDeRn (2013c): MoDeRn Case Studies Report. Deliverable D4.1

MoDeRn (2013d): Monitoring During the Staged Implementation of Geological Disposal: Technology Summary Report. Deliverable D

Nowotny, H. (2003): Democratising expertise and socially robust knowledge. Science and Public Policy 30(3), S. 151–156

US NRC – US Nuclear Regulatory Commission (2012): Disposal of High-Level Radioactive Wastes in Geologic Repositories – Performance Confirmation Program. Code of Federal Regulations 10, Part 60, Subpart F. https://www.govinfo.gov/content/pkg/CFR-2018-title10-vol2/xml/CFR-2018-title10-vol2-sec60-140.xml [Zugriff am 01.12.2021]

Wilsdon, J.; Willis, R. (2004): See-through science: Why public engagement needs to move upstream. Project Report, London

Wynne, B. (1989): Sheepfarming after Chernobyl: A case study in communicating scientific information. Environment Science and Policy for Sustainable Development 31(2), S. 10–39

Anne Eckhardt

Entsorgung unter Kontrolle? Erfahrungen mit dem Monitoring von Tiefenlagern in der Schweiz

In der Schweiz wurde bereits 2003 ein Tiefenlagerkonzept mit Vorkehrungen zur Überwachung rechtlich verankert. Aus der Entwicklung dieses Konzepts und den bisherigen Erfahrungen mit seiner Umsetzung lassen sich Folgerungen ziehen, die für die Realisierung der Tiefenlagerung mit Vorkehrungen für Monitoring und Rückholbarkeit international, insbesondere in Deutschland, von Interesse sein können.

1 Entstehung des Konzepts der geologischen Tiefenlagerung

Die Schweiz verfolgt bei der Entsorgung hochradioaktiver Abfälle ein spezifisches Konzept: das Konzept der geologischen Tiefenlagerung (vgl. Abb. 1). Sowohl schwach- und mittelradioaktive als auch hochradioaktive Abfälle sollen in einer bergwerksartigen Anlage in tiefen geologischen Formationen eingelagert werden. Diese Anlage wird letztlich in ein Endlager überführt. Vor dem Verschluss des Tiefenlagers ist jedoch, anders als in Endlagerkonzepten, eine Beobachtungsphase vorgesehen. Die Dauer dieser Phase ist nicht vorbestimmt. Im Verlauf der Beobachtungsphase wird das Verhalten der Abfälle, der Verfüllung und des Wirtgesteins überwacht. Damit die Sicherheit des Lagers durch die Beobachtung nicht beeinträchtigt wird, findet die Überwachung in einem sogenannten Pilotlager statt. Das Pilotlager enthält eine repräsentative kleine Menge von Abfällen. Räumlich und hydraulisch (d. h. bezogen auf Flüssigkeitsbewegungen) ist es vom Hauptlager getrennt, wo der weit überwiegende Teil der Abfälle eingelagert ist.

Das Konzept der geologischen Tiefenlagerung wurde Ende der 1990er Jahre entwickelt. Vorausgegangen waren einige Bestrebungen, die Entsorgung der radioaktiven Abfälle dauerhaft zu lösen, die letztlich jedoch nicht zum Ziel geführt hatten:

Abbildung 1: Schematische Darstellung des von der EKRA entwickelten Lagerkonzepts

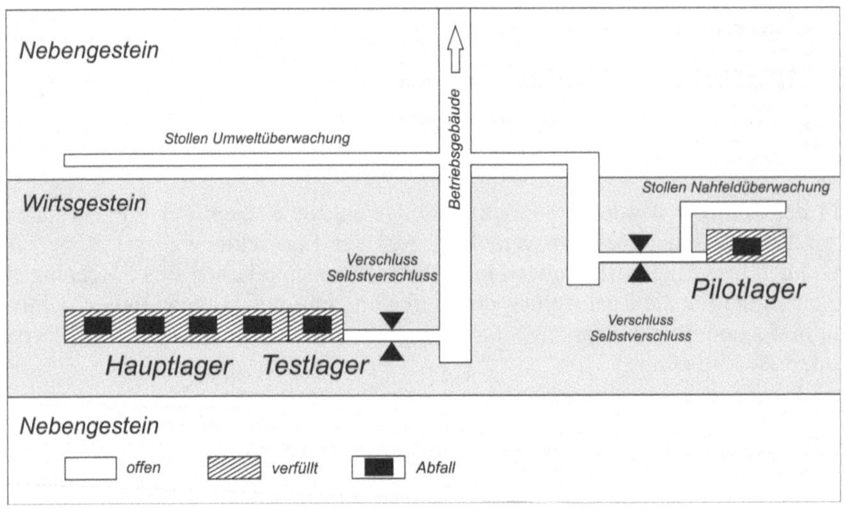

Quelle: angelehnt an EKRA 2000, S. 56

1994 hatten die Abfallverursacher, vertreten durch die Genossenschaft für nukleare Entsorgung Wellenberg (GNW), ein Gesuch um Rahmenbewilligung für ein Endlager für schwach- und mittelaktive Abfälle gestellt. Als Standort war der Wellenberg im Kanton Nidwalden vorgesehen. 1995 lehnte die Bevölkerung des Kantons Nidwalden wesentliche Voraussetzungen für die Erteilung des Rahmenbewilligungsgesuches ab (Hadermann et al. 2014, S. 139 f.).

1998 hatte sich die Arbeitsgruppe „Energie-Dialog Entsorgung“ im Auftrag des zuständigen Bundesrats, Moritz Leuenberger, mit Grundsatzfragen der nuklearen Entsorgung befasst. Dabei zeigte sich, dass die Betreiber von Kernanlagen und die Umweltorganisationen zwei grundsätzlich verschiedene Lagerkonzepte bevorzugten. Der Vorsitzende der Arbeitsgruppe, der Philosoph Hans Ruh, empfahl, das Konzept der kontrollierten und rückholbaren Langzeitlagerung, das wichtige Umweltorganisationen favorisierten, vertiefter zu untersuchen. Die Betreiberseite trat dagegen für das Konzept der Endlagerung ein (Ruh 1998, S. 13).

Anfang 1999 erbrachten Gespräche zur Befristung des Betriebs der bestehenden Kernkraftwerke und zur Lösung des Entsorgungsproblems zwischen dem Bundesrat sowie den Standortkantonen, Umweltorganisationen und den Betreibern der Kernkraftwerke kein zufriedenstellendes Ergebnis. Daraufhin setzte der Vorsteher des Eidgenössischen Departements für Umwelt, Verkehr, Energie und Kommunikation (UVEK) im Juni 1999 die „Expertengruppe Entsorgungskonzepte für radioaktive Abfälle“ (EKRA) ein. Die Expertengruppe war unter anderem damit beauftragt, Grundlagen für einen Vergleich verschiedener Entsorgungskonzepte für radioaktive Abfälle zu erarbeiten (EKRA 2000, S. 1). Sie entwickelte zwischen Juni 1999 und Januar 2000 das neue Konzept der geologischen Tiefenlagerung. Nach einer längeren Phase der Stagnation fand dieses Konzept breite gesellschaftliche und politische Unterstützung. 2003 wurde es in der neuen Kernenergiegesetzgebung der Schweiz verankert (KEG 2003; KEV 2004).

Das Konzept der geologischen Tiefenlagerung verdankt seinen Erfolg vor allem der Tatsache, dass es mit ihm gelang, sowohl wesentliche Positionen und Anliegen der Befürworter der Endlagerung als auch wesentliche Positionen und Anliegen der Befürworter der kontrollierten Langzeitlagerung aufzunehmen und in einem gemeinsamen Lösungsansatz zu vereinen. Ein wichtiges Argument der Befürworter eines Endlagers war beispielsweise, dass nur eine passiv sichere Entsorgungsanlage dauerhafte Sicherheit bieten könne. Dies gelte insbesondere, weil das Mehrfachbarrierensystem eines Endlagers auch im Fall gesellschaftlicher Krisen Schutz vor den Auswirkungen radioaktiver Abfälle auf Menschen und Umwelt biete. Die Befürworter der kontrollierten geologischen Langzeitlagerung dagegen setzten mehr Vertrauen in die dauerhafte Gewährleistung der Sicherheit durch Menschen, die die Entsorgungsanlage aktiv bewirtschaften. Diese Menschen sollten gegenwärtig und in der Zukunft in der Lage sein, auf neue, ggf. unerwartete Entwicklungen schnell zu reagieren und Fortschritte aufzunehmen, die zu einer verbesserten Sicherheit beitragen. In ihrem Vorschlag verband die EKRA dann die Feststellung, dass nach aktuellem Wissensstand allein die Endlagerung Sicherheit über sehr lange Zeiträume gewährleisten könne – mit Offenheit für einen möglichen Sicherheitszuwachs durch Erkenntnisgewinn und technischen Fortschritt in den kommenden Jahrzehnten bis Jahrhunderten (EKRA 2000, S. 71 f.). In Tabelle 1 sind explizite und implizite Positionen und Anliegen beider Seiten[1] sowie die entsprechenden Lösungsansätze der Expertengruppe EKRA dargestellt:

Die obersten Schutzziele der Entsorgung radioaktiver Abfälle waren sowohl bei den Befürwortern der kontrollierten Langzeitlagerung als auch bei den Befürwortern der Endlagerung unbestritten. Es handelt sich dabei um:

1 Vgl. z. B. Ruh (1998, S. 9–13); zu den Lösungsansätzen siehe EKRA (2000).

- Sicherheit: Schutz von Menschen und Umwelt vor schädigenden Einwirkungen der Abfälle
- Gerechtigkeit: Gewährleistung der Handlungsfreiheit künftiger Generationen

Tabelle 1: Integration unterschiedlicher Positionen im Konzept der geologischen Tiefenlagerung durch die Expertengruppe EKRA

Positionen und Anliegen der Befürworter eines Endlagers	*Positionen und Anliegen der Befürworter eines kontrollierten Langzeitlagers*	*Vorschlag der Expertengruppe EKRA*
Passive Sicherheit muss im Vordergrund stehen.	Aktive Kontrolle ist wesentlich.	Gewährleistung der passiven Sicherheit im Hauptlager bald nach Einlagerung Gewährleistung von aktiver Kontrolle während der Beobachtungsphase im Pilotlager
Eine erleichterte Rückholbarkeit ist nach dem Verschluss des Lagers nicht erforderlich und wirkt sich voraussichtlich kontraproduktiv auf die Sicherheit des Lagers aus.	Die Rückholbarkeit der Abfälle muss gewährleistet sein, ohne dass dabei die Schutzziele infrage gestellt werden.	Verbindung von Endlagerung und Reversibilität
Das Lager muss Schutz gegenüber ungewissen Risiken in der Zukunft bieten.	Das Lager muss es erlauben, von Fortschritten in der Zukunft zu profitieren.	Verbindung von Endlagerung und Reversibilität Zeitlich nicht befristete Beobachtungsphase nach der Einlagerung der Abfälle Schnellverschluss der gesamten Anlage in Krisenfällen
Die Fachleute, die gegenwärtig an der Entsorgung arbeiten, sind glaubwürdig.	Die Fachleute, die gegenwärtig an der Entsorgung arbeiten, sind nicht glaubwürdig.	Überprüfung der Sicherheit der gesamten Anlage am Pilotlager während zeitlich nicht befristeter Beobachtungsphase
Das verfügbare Fachwissen reicht aus, um die Entsorgung in die Hand zu nehmen.	Das verfügbare Fachwissen reicht nicht aus, um die Entsorgung in die Hand zu nehmen.	Überprüfung der Sicherheit der gesamten Anlage am Pilotlager während zeitlich nicht befristeter Beobachtungsphase Verbindung von Endlagerung und Reversibilität
Künftige Generationen sollen von Verpflichtungen, für die Sicherheit des Lagers zu sorgen, entlastet werden.	Künftige Generationen sollen in der Lage sein, selbst zu entscheiden, wie sie mit den Abfällen umgehen wollen.	Verbindung von Endlagerung und Reversibilität

Quelle: Eigene Darstellung

Die Expertengruppe EKRA wurde von dem Genfer Geologen Walter Wildi geleitet. Sie umfasste sieben Mitglieder und wurde von einem wissenschaftlichen Sekretär unterstützt. Vier der Mitglieder waren Geologen, ein Mitglied Bauingenieur und ein Mitglied Biologin; der wissenschaftliche Sekretär der Gruppe war Chemiker. Die Geisteswissenschaften waren lediglich mit einer Person, einem Theologen und Ethiker, vertreten.

Welche Aspekte ein Sozialwissenschaftler oder eine Sozialwissenschaftlerin in die Arbeit der Gruppe eingebracht hätte, darüber lässt sich im Nachhinein nur spekulieren. Vorstellbar ist, dass sich die Arbeitsgruppe eingehender mit den unterschiedlichen Vorstellungen von Sicherheit, Gerechtigkeit und Kontrollierbarkeit der Befürworter der Endlagerung und der Befürworter der kontrollierten Langzeitlagerung auseinandergesetzt hätte. Naheliegend ist, dass sie organisatorische Aspekte des Entsorgungskonzepts stärker in ihre Arbeit einbezogen hätte, wie etwa Gouvernanz-Fragen und menschliche Einflüsse auf die Sicherheit. Tatsächlich publizierte die EKRA im Jahr 2002 noch einen „Beitrag zur Entsorgungsstrategie für die radioaktiven Abfälle in der Schweiz“ (EKRA 2002). Dieser Bericht, der kurz und allgemein gehalten war, fand jedoch kaum Resonanz.

Ungeachtet der Tatsache, dass die EKRA hauptsächlich aus Natur- und Ingenieurwissenschaftlern bestand, lässt ihr Bericht Offenheit gegenüber sozialwissenschaftlichen Fragestellungen erkennen. Die gesellschaftliche „Akzeptanz“ eines Entsorgungskonzepts beispielsweise ist im Schlussbericht der Arbeitsgruppe ein wiederkehrendes Thema (vgl. z. B. EKRA 2000, S. 2). Zudem ist im Bericht der EKRA das Bestreben feststellbar, konsequent sowohl den Anliegen der Befürworter der kontrollierten Langzeitlagerung als auch jenen der Endlagerung entgegenzukommen. Diese Offenheit war eine wesentliche Voraussetzung für den Erfolg, den das Konzept der geologischen Tiefenlagerung in der Schweiz bis heute hat.

2 Umsetzung des Konzepts

Die Empfehlungen der EKRA waren allgemein und konzeptionell formuliert. Dies entsprach dem Auftrag, den die EKRA erhalten hatte, und wäre angesichts des knappen Zeitrahmens, in dem das Konzept entwickelt wurde, auch kaum anders machbar gewesen. So enthält der Bericht der EKRA beispielsweise keine genaueren Überlegungen dazu, wie die empfohlene zeitlich nicht limitierte Beobachtungsphase unter den Aspekten von Sicherheit und Machbarkeit umgesetzt werden kann und soll. Die EKRA weist aber darauf hin, dass detaillierte Untersuchungen zu den Einflüssen der Beobachtungsphase auf die Langzeitsicherheit erforderlich seien (EKRA 2000, S. 67).

Das Konzept der geologischen Tiefenlagerung war im Dialog mit wichtigen Akteuren wie Abfallverursachern und Umweltorganisationen entwickelt worden. In den folgenden Jahren wurde dieses Konzept in einem demokratischen Prozess von Parlament (Bundesversammlung der Schweizerischen Eidgenossenschaft) und Regierung (Schweizerischer Bundesrat) in rechtliche Vorgaben umgesetzt. Seither liegt die weitere Ausgestaltung und Umsetzung des Konzepts vor allem bei natur- und ingenieurwissenschaftlich ausgerichteten Spezialist*innen. Diese natur- und ingenieurwissenschaftlich ausgerichteten Spezialist*innen sind aufseiten der Abfallverursacher, vertreten durch die Nationale Genossenschaft für die Lagerung radioaktiver Abfälle (Nagra), für die Realisierung einer sicheren Entsorgung verantwortlich. Aufseiten der Aufsichtsbehörde, dem Eidgenössischen Nuklearsicherheitsinspektorat (ENSI), konkretisieren sie die rechtlichen Vorgaben (ENSI 2020) und überwachen die Einhaltung der rechtlichen Vorschriften. Durch die partizipativen Elemente im Standortauswahlverfahren, dem Sachplanverfahren geologische Tiefenlager (BFE 2011), ist eine gewisse Verbindung zur interessierten Öffentlichkeit bei der Konzeption der Tiefenlagerung sichergestellt. Durch die Zusammenarbeit des ENSI mit dem Bundesamt für Energie (BFE), das die Federführung des Sachplanverfahrens übernommen hat, werden fallweise auch sozialwissenschaftliche Kompetenzen in die Umsetzung der geologischen Tiefenlagerung eingebunden.

Das Konzept der geologischen Tiefenlagerung vereinigte also Bedürfnisse in der Gesellschaft mit den Ansichten von Spezialist*innen aus Natur- und Ingenieurwissenschaften. Nachdem diese Offenheit gegenüber zivilgesellschaftlichen Interessen zum politischen Erfolg geführt hatte, wurde die Konkretisierung und Umsetzung des Konzepts weitgehend an die Gemeinschaft der Spezialist*innen aus Natur- und Ingenieurwissenschaften zurückgegeben. Dort haben dann die charakteristischen Sichtweisen dieser Gemeinschaft Spuren hinterlassen. So wurde beispielsweise die ursprüngliche Forderung nach „Kontrollierbarkeit" des Lagers (Ruh 1998) im Wesentlichen in die Forderung nach einer systematischen Überwachung und nach der Möglichkeit der „Rückholung ohne grossen Aufwand" (KEG 2003) der radioaktiven Abfälle bis zum Ende der Beobachtungsphase übersetzt. Ob Kontrollierbarkeit beispielsweise auch gesellschaftliche Teilhabe an der Überwachung der Entsorgungsanlage bedeutet, wurde nicht geregelt. Die EKRA hatte im Jahr 2000 formuliert, das Pilotlager solle als „Nachweislager" dienen, „welches über den Verschluss des Hauptlagers hinaus eine Langzeitkontrolle ermöglicht" (EKRA 2000, S. 53). In Art. 66 der Kernenergieverordnung ist seit 2004 festgehalten, im Pilotlager seien „das Verhalten der Abfälle, der Verfüllung und des Wirtgesteins bis zum Ablauf der Beobachtungsphase zu überwachen. Bei der Überwachung sind im Hinblick auf den Verschluss Daten zur Erhärtung des Sicherheitsnachweises zu ermitteln". Ob etwa die gesellschaftlichen Anforderungen nach Langzeitkontrolle durch Erhärtung des Sicherheitsnachweises erfüllt werden (Röhlig/Eckhardt 2017), bleibt dabei offen.[2]

Die bisherigen Erfahrungen legen daher nahe: Die Umsetzung einer Entsorgungsoption, die sowohl an ingenieur- und naturwissenschaftlichem Fachwissen als auch an Bedürfnissen innerhalb der Gesellschaft ausgerichtet ist, sollte sowohl von Vertretern verschiedener Fachdisziplinen als auch von Vertretern der Zivilgesellschaft durchgeführt oder zumindest begleitet werden. Bei der Umsetzung der Entsorgungsoption müssen alle Beteiligten daran arbeiten, ein generatives System aufzubauen und zu erhalten (vgl. Oliver Sträter in diesem Band), das den eigenen Weg permanent hinterfragt, Selbstzufriedenheit vermeidet und offen für neue Ideen ist.

3 Pilotlager: Gewinn oder Gefahr für die Sicherheit?

Eine Besonderheit der schweizerischen geologischen Tiefenlagerung ist das Pilotlager, also jener Teil der untertägigen Anlage, in der ein repräsentatives Abfallinventar über den Verschluss des Hauptlagers hinaus überwacht werden soll. Die Expertengruppe EKRA präsentierte in ihrem Schlussbericht Vorstellungen zur Ausgestaltung des Pilotlagers, die jedoch noch vage blieben. So machte die EKRA beispielsweise keine Aussagen dazu, wie das repräsentative Inventar im Pilotlager nach Beendigung der Pilotphase entsorgt werden solle oder welche Anforderungen an den Selbstverschluss (EKRA 2000, S. 53 f.) des Lagers zu stellen seien. Sie wies allerdings auch selbst darauf hin, dass zur Überführung in den Endlagerzustand noch Abklärungen erforderlich seien (EKRA 2000, S. 3). Auch die wichtige Frage, wie repräsentativ die Beobachtungen im Pilotlager tatsächlich für die Vorgänge im Hauptlager sind, wurde von der EKRA nicht eingehender diskutiert. Haupt- und Pilotlager sind räumlich und hydraulisch voneinander getrennt und befinden sich demnach in einem ähnlichen, nicht aber im gleichen geologischen Umfeld. Das Pilotlager beinhaltet zwar einen „kleinen, aber repräsentativen Anteil der Abfälle" (EKRA 2000, S. 2), ermöglicht es jedoch nicht, das Hauptlager tatsächlich lückenlos zu überwachen. Sollte beispielsweise ein geringer Teil der Endlagerbehälter Fabrikationsmängel aufweisen, würden diese Mängel im Pilotlager nicht zwangsläufig aufgedeckt. International hat das schweizerische Konzept denn auch bisher keine Nachahmer gefunden, obwohl Monitoring und Rückholbarkeit mittlerweile

2 Zur Einschätzung des Schweizer Verfahrens siehe z. B. auch Streffer et al. (2011, S. 396–404) und Kuppler (2017).

in vielen Ländern, zum Beispiel in Deutschland, intensiv diskutiert werden (vgl. Endlagerkommission 2016 und Grunwald in diesem Band).

Das Monitoring eines Tiefenlagers ist heute international vor allem darauf ausgerichtet, Entwicklungen, die die Sicherheit des Lagers beeinträchtigen könnten, frühzeitig zu entdecken. Damit soll es ermöglicht werden, eventuelle Schäden zu verhindern oder in Grenzen zu halten, zum Beispiel indem Abfälle rückgeholt und neu verpackt werden. Monitoring wird also vorgesehen, um einen Beitrag zur Verbesserung der Sicherheit des Tiefenlagers zu leisten. Gleichzeitig ist jedoch zu erwarten, dass Monitoring die Sicherheit eines Tiefenlagers in vorhersehbarer Art und Weise beeinträchtigt.

Dies gilt zum einen für die Langzeitsicherheit: Ein klassisches Endlager folgt einem einfachen, überschaubaren Konzept und wird daher heute als robuster gegenüber ungewissen Entwicklungen in der Zukunft angesehen als das komplizierter aufgebaute geologische Tiefenlager mit Hauptlager und Pilotlager. Monitoring kann die Sicherheitsbarrieren des Lagers schwächen, indem es beispielsweise neue Wegsamkeiten für Radionuklide schafft und Einwirkungen von außen, wie das Eindringen von Wasser, begünstigt. Die verlängerte Offenhaltung, Emissionen und weitere Beanspruchungen während der Beobachtungsphase können zu Schäden an den Sicherheitsbarrieren führen (Eckhardt/Rippe 2016, S. 72 f.). Art. 11 der schweizerischen Kernenergieverordnung zielt daher darauf ab, Beeinträchtigungen der Langzeitsicherheit zu vermeiden. Demnach dürfen „Vorkehrungen zur Erleichterung von Überwachung und Reparaturen des Lagers oder zur Rückholung der Abfälle die passiven Sicherheitsbarrieren nach dem Verschluss des Lagers nicht beeinträchtigen" (KEV 2017).

Neben der Langzeitsicherheit ist auch die Betriebssicherheit des Pilotlagers zu beachten. Während des Betriebs des Pilotlagers, der sich zeitlich mit dem Einlagerungsbetrieb in das Hauptlager überschneidet und auch die anschließende Beobachtungsphase umfasst, sind vor allem Personen Risiken ausgesetzt, die für den Betrieb und Unterhalt des Pilotlagers und seiner Zugänge verantwortlich sind, das Monitoring vornehmen oder das Lager gegen Einwirkungen Dritter sichern. Auch der Bau des Pilotlagers, das zusätzlich zum Hauptlager aufgefahren wird, ist mit Risiken für die an diesen Arbeiten beteiligten Personen verbunden, zum Beispiel aufgrund von Arbeitsunfällen, Staub- und Lärmemissionen, körperlicher Belastung etc. Hinzu kommen Risiken durch Einwirkungen Dritter während der verlängerten Offenhaltung des Lagers und potenzielle Störfälle während des Baus des Pilotlagers und dessen Betrieb. Alle diese Risiken lassen sich aus heutiger Sicht relativ gut einschätzen – falls man davon ausgeht, dass auch in Zukunft ähnliche Sicherheitsstandards wie zum gegenwärtigen Zeitpunkt gelten und ähnliche Techniken zum Einsatz kommen.

Nicht einschätzbar ist dagegen der Sicherheitsgewinn, der aus dem Monitoring resultiert. Dies ist im Wesentlichen darauf zurückzuführen, dass das

Monitoring vor allem als Vorsorge gegenüber Risiken gedacht ist, die bis zur Beendigung der Einlagerung ins Hauptlager noch nicht bekannt und daher ungewiss sind. Die Risiken, die durch das Monitoring bedingt sind, lassen sich aus heutiger Perspektive also einschätzen, nicht aber der Sicherheitsgewinn, der mit dem Monitoring erzielt wird. Daher ist es auch nicht möglich, die zusätzlichen Risiken und den erzielten Sicherheitsgewinn direkt gegeneinander abzuwägen. Die Frage, ob nach dem Verschluss eines Tiefenlagers weiterhin Monitoring betrieben werden soll und, wenn ja, in welcher Form und wie lange, muss daher letztlich politisch entschieden werden. In der Schweiz hat sich die Bundesversammlung 2003 für das Konzept der geologischen Tiefenlagerung entschieden, das ein Pilotlager zum Zweck des Monitorings einschließt. Die Entscheidung darüber, wie lange die Beobachtungsphase andauern soll, hat sie künftigen Entscheidungsträgern überlassen.

4 Auf dem Weg zu einem erweiterten Verständnis von Monitoring

Anzumerken ist, dass sich das Monitoring eines geologischen Tiefenlagers in der Schweiz nicht nur auf das Pilotlager beschränken wird. Bereits während der Erkundung des künftigen Lagerstandorts und vor Baubeginn wird ein Umweltmonitoring vorgenommen, das auch nach dem Verschluss des Lagers weitergeführt werden kann. Monitoring findet voraussichtlich während des Baus des Tiefenlagers und der Einlagerung der Abfälle statt. Monitoring wird zudem in den Testbereichen durchgeführt, einem Felslabor, das am Standort des geplanten Tiefenlagers errichtet wird und unter anderem dazu dient, die sicherheitsrelevanten Eigenschaften des Wirtgesteins standortspezifisch genauer abzuklären. Neben dem Monitoring, das auf potenzielle Schäden durch die Freisetzung von Radionukliden ausgerichtet ist, ist auch ein Monitoring zu den nicht radiologisch bedingten Umweltauswirkungen und zu Veränderungen im gesellschaftlichen Umfeld sinnvoll. Im Schweizer Sachplanverfahren wurden daher Abklärungen zu den sozioökonomischen und ökologischen Auswirkungen eines geologischen Tiefenlagers vorgenommen (vgl. z. B. BFE 2015).

International wird gegenwärtig diskutiert, inwieweit ein Monitoring nicht nur von naturwissenschaftlich-technischen Erwägungen, sondern auch von den Bedürfnissen der von einer Entsorgungsanlage Betroffenen geleitet werden soll. Das Forum on Stakeholder Confidence der NEA ist bestrebt, die Betroffenen zu Beteiligten (*partnership*) oder sogar zu Eignern (*ownership*) bei der Entsorgung radioaktiver Abfälle zu machen (OECD/NEA 2013). Dies spricht für eine starke und aktive Rolle der Betroffenen beim Monitoring. Wesentlich ist zudem, frühzeitig Vorstellungen dazu zu entwickeln, wie mit den Ergebnissen des Monitorings umgegangen werden soll. Die Entscheidung, ob ein Messwert, der vom Erwarteten abweicht, als Messfehler interpretiert wird oder weitergehende Abklärungen auslöst, kann weitreichende Konsequenzen nach sich ziehen. Daher sollte frühzeitig über ein Kontroll- und Entscheidungsverfahren nachgedacht werden, an dem Regierungsorganisationen, unabhängige Experten und die interessierte Öffentlichkeit beteiligt werden könnten. Mit seinen partizipativen Elementen ebnet das Sachplanverfahren in der Schweiz den Weg zu einem derartigen Verfahren (Kuppler/Hocke 2012).

Die genaueren Anforderungen an das Monitoring im Pilotlager müssen in den kommenden Jahren konkretisiert werden. Die Sicherheitsbehörde ENSI trifft daher Abklärungen zum Monitoring. Die Nagra, die für die sichere Entsorgung verantwortlich ist, arbeitet international vernetzt an konkreten Lösungsansätzen. Sowohl das ENSI als auch die Nagra sind in erster Linie auf die Beantwortung von naturwissenschaftlich-technischen Sicherheitsfragen ausgerichtet. Handlungsbedarf besteht jedoch auch unter sozialwissenschaftlichen Gesichtspunkten. Dort geht es beispielsweise darum, das gesellschaftliche Bedürfnis nach „Kontrollierbarkeit" eines Lagers besser zu verstehen, Veränderungen bei diesem Bedürfnis frühzeitig zu erfassen und Vorstellungen zur Umsetzung des Schutzziels zur Gerechtigkeit zu entwickeln. Obwohl das Konzept der geologischen Tiefenlagerung seit zehn Jahren in der Kernenergiegesetzgebung verankert war, waren 2013 nur 50% der Schweizer Bevölkerung der Meinung, dass die Entsorgung radioaktiver Abfälle im tiefen Untergrund die beste Lösung darstelle. 82% der Schweizer Bevölkerung stimmten das Aussage zu, es gebe keinen sicheren Weg, radioaktive Abfälle zu entsorgen (BFE 2013, S. 5).

Letztlich stellt sich damit die grundsätzliche Frage, welche Ziele mit dem Monitoring im Pilotlager erreicht werden sollen. Geht es in erster Linie darum, mehr Schutz vor ungewissen Risiken in der Zukunft zu erreichen? Steht die Entscheidungsunterstützung für den endgültigen Verschluss des geologischen Tiefenlagers im Vordergrund? Oder soll vor allem Vertrauen gebildet und sollen Betroffene aktiv in Entscheidungen eingebunden werden, die ihre eigene Sicherheit beeinflussen? Kann das Monitoring auch zum dauerhaften Kompetenzerhalt auf dem Gebiet der Entsorgung genutzt werden? Ist wissenschaftlicher Erkenntnisgewinn ebenfalls ein legitimes Ziel? Um diese und ähnliche Fragen wird in den kommenden Jahren in der Schweiz und international noch eine intensive Diskussion geführt werden müssen.

Literatur

BFE – Bundesamt für Energie (2011): Sachplan geologische Tiefenlager. Konzeptteil. 2. April 2008 (Revision vom 30. November 2011). Bern

BFE – Bundesamt für Energie (2013): Attitudes towards radioactive waste in Switzerland. Report. TNS opinion im Auftrag des Bundesamts für Energie. Bern

BFE – Bundesamt für Energie (2015): Die sozioökonomischen und ökologischen Auswirkungen eines geologischen Tiefenlagers auf die Standortregion Zürich Nordost. Synthesebericht zur SÖW, den Zusatzfragen und der Gesellschaftsstudie. Bundesamt für Energie. Bern

Eckhardt, A.; Rippe, K.P. (2016): Risiko und Ungewissheit bei der Entsorgung hochradioaktiver Abfälle. vdf Hochschulverlag AG, Zürich

EKRA – Expertengruppe Entsorgungskonzepte für radioaktive Abfälle (2000): Entsorgungskonzepte für radioaktive Abfälle. Schlussbericht. Bern

EKRA – Expertengruppe Entsorgungskonzepte für radioaktive Abfälle (2002): Beitrag zur Entsorgungsstrategie für die radioaktiven Abfälle in der Schweiz. Bern

Endlagerkommission (2016): Verantwortung für die Zukunft. Ein faires und transparentes Verfahren für die Auswahl eines nationalen Endlagerstandortes. Abschlussbericht der Kommission Lagerung hoch radioaktiver Abfallstoffe. Berlin

ENSI – Eidgenössisches Nuklearsicherheitsinspektorat (2020): Geologische Tiefenlager. Richtlinie für die schweizerischen Kernanlagen ENSI-G03/d. Brugg

Hadermann, J.; Issler, H.; Zurkinden, A. (2014): Die nukleare Entsorgung in der Schweiz 1945–2006. Zürich

KEG – Kernenergiegesetz vom 21. März 2003 (Stand am 1. Januar 2020). SR 732.1

KEV – Kernenergieverordnung vom 10. Dezember 2004 (Stand am 1. Februar 2019). SR 732.11

Kuppler, S. (2017): Effekte deliberativer Ereignisse in der Endlagerpolitik. Deutschland und die Schweiz im Vergleich von 2001 bis 2010. Wiesbaden

Kuppler, S.; Hocke, P. (2012): Monitoring in einem Pilotlager. Kontrollierte Deponierung von Nuklearabfällen im Konzept eines Schweizer Tiefenlagers. Technikfolgenabschätzung – Theorie und Praxis 21(3), S. 43–51

OECD; NEA – Organisation for Economic Co-operation and Development; Nuclear Energy Agency (2013): Stakeholder confidence in radioactive waste management. An annotated glossary of key terms. Paris

Röhlig, K.-J.; Eckhardt, A. (2017): Primat der Sicherheit. Ja, aber welche Sicherheit ist gemeint? GAIA 2, S. 103–105

Ruh, H. (1998): Energie-Dialog Entsorgung. Schlussbericht des Vorsitzenden zu Handen des Eidg. Departements für Umwelt, Verkehr, Energie und Kommunikation. Bern

Streffer, C.; Gethmann, C.F.; Kamp, G.; Kröger, W.; Rehbinder, E.; Renn, O.; Röhlig, K.-J. (2011): Radioactive Waste. Technical and Normative Aspects of its Disposal. Berlin/Heidelberg

Karl-Heinz Lux, Ralf Wolters, Juan Zhao

Direktes Monitoring des Tiefenlagerverhaltens mithilfe von Überwachungssohle und Messbohrlöchern. Ein Beitrag zur Vermittlung zwischen Technik und Gesellschaft?

1 Motivation für das direkte Monitoring eines Tiefenlagers

Im Zusammenhang mit der Entsorgung hochradioaktiver Abfälle in tiefen geologischen Formationen in Deutschland sind bereits im Jahr 2010 in den vom Bundesministerium für Umwelt, Naturschutz und Reaktorsicherheit veröffentlichten Sicherheitsanforderungen an die Endlagerung wärmeentwickelnder radioaktiver Abfälle (BMU 2010) zunächst während der Einlagerung eine planerisch berücksichtigte Rückholbarkeit der wärmeentwickelnden hochradioaktiven Abfälle und anschließend für einen Zeitraum von 500 Jahren nach Verschluss des Endlagers eine grundsätzliche Bergbarkeit gefordert worden. Als Entscheidungsgrundlage für eine Rückholung oder eine Bergung der Abfälle aus der Entsorgungsanlage[1] ist ein geeignetes Monitoringkonzept zur Überwachung des unmittelbaren Ablagerungsbereiches sowohl bezüglich Radionuklidmobilisierung und Behälterintegrität als auch bezüglich des Tiefenlagersystemverhaltens notwendig. Ein derartiges Monitoringkonzept muss schon im Rahmen des Standortauswahlprozesses in die konfigurative Planung des Entsorgungsanlagensystems implementiert werden, da die konkrete Ausgestaltung des Monitoringkonzepts Einfluss haben kann, z. B. auf den untertägigen Raumbedarf zur Errichtung des Tiefenlagers/Endlagers.

Zur Überwachung der Entsorgungsanlage während der Einlagerungsphase wie auch noch einige Zeit danach wird im Rahmen des Forschungsprojekts ENTRIA (Entsorgungsoptionen für radioaktive Reststoffe: Interdisziplinäre Analysen und Entwicklung von Bewertungsgrundlagen) vorgeschlagen, ein direktes Monitoring der versetzten Einlagerungssohle in das Entsorgungsanlagenkonzept zu implementieren (Stahlmann et al. 2015). Dieses direkte Monitoring könnte z. B. durch Anordnung einer vor Einlagerungsbeginn errichteten und auch noch nach Versatz und Verschluss der Einlagerungssohle längerfristig

1 Im vorliegenden Beitrag wird die Entsorgungsanlage bis zum Ende der Beobachtungsphase als Tiefenlager und danach als Endlager bezeichnet.

bis zum Ende der in ihrem zeitlichen Rahmen später noch festzulegenden Monitoringphase offen zu haltenden Überwachungssohle erfolgen. Diese Überwachungssohle wird oberhalb der Einlagerungssohle angeordnet und mit der Einlagerungssohle über Beobachtungs- bzw. Messbohrlöcher verbunden. Aus dieser Konzeption resultiert konfigurativ ein zweisöhliges Entsorgungsbergwerk. Die Beobachtungs- bzw. Messbohrlöcher sollten zum Ausschluss einer potenziellen Strahlenbelastung in der Überwachungssohle mit einem beweglichen Verschlusssystem ausgestattet werden.

Im Rahmen des Forschungsprojekts ENTRIA wird die Möglichkeit eines auch über die Einlagerungsphase hinausreichenden direkten Monitorings in einem generischen, d. h. idealisierten und nicht standortbezogenen, zweisöhligen Tiefenlager-/Endlagersystem für unterschiedliche Wirtsgesteinsformationen untersucht – einerseits aus geotechnischer Sicht (Stahlmann et al. 2015), andererseits im Hinblick auf die Gewährleistung der Langzeitsicherheit (Lux et al. 2017a-d). Vor diesem Hintergrund stehen dabei insbesondere das fluiddynamische Systemverhalten als Träger der Radionuklidmigration und messtechnisch beobachtbare Zustandsgrößenentwicklungen innerhalb des Nahfeldes der Einlagerungssohle als mögliche Monitoringparameter im Vordergrund. Der Simulationszeitraum umfasst dabei die sukzessive Auffahrung des Tiefenlagers und der Überwachungssohle mit den Messbohrlöchern, die Abfalleinlagerung, den begleitenden Versatz und den Verschluss der Einlagerungssohle und die nachfolgende weitere Beobachtungsphase sowie dann auch den Versatz und Verschluss der Überwachungssohle und der Schächte, die Überführung des Tiefenlagers in ein Endlager sowie die weitere Endlagerentwicklung bis zu einer Million Jahre.

Das im Rahmen des Forschungsprojekts ENTRIA erarbeitete Monitoringkonzept sowie dazugehörige geotechnische und fluiddynamische Analysen sind bereits umfassend vorgestellt worden (Lux et al. 2017a-d). Der vorliegende Beitrag beruht auf diesen Ausführungen und ist hier auf die Grundgedanken des Konzepts und seiner Umsetzung fokussiert.

2 Möglichkeiten zur konfigurativen Ausgestaltung von Monitoringkonzepten

Im Schlussbericht der Kommission „Lagerung hoch radioaktiver Abfallstoffe" (auch verkürzt Endlagerkommission genannt), die im Jahr 2014 zur Überprüfung des Standortauswahlgesetzes eingesetzt worden ist, wird in Abschnitt 3.3 / Etappe 4 noch vorerst nur sehr grundsätzlich ohne Hinweise zur technischen Umsetzbarkeit oder sicherheitstechnischen Notwendigkeit auch eine Beobachtung des Endlagerverhaltens nach Ende des Einlagerungsbetriebs angesprochen (Kommission „Lagerung hoch radioaktiver Abfallstoffe“ 2016).

Erste Ansätze zur Implementierung derartiger Überlegungen in ein Endlagerkonzept laufen z. B. in der Schweiz darauf hinaus, das dort als geologisches Tiefenlager bezeichnete Endlager nach Ende der Einlagerungsphase durch ein sogenanntes Pilotlager, das außerhalb der Einlagerungsbereiche des Tiefenlagers eingerichtet ist, in seinem grundsätzlichen Verhalten längerfristig zu überwachen (Nagra 2014, vgl. den Beitrag von Eckhardt in diesem Band). Diese Überwachung ist allerdings als nur mittelbar anzusehen, da im Pilotlager zwar standortbezogen die ablaufenden multiphysikalischen Prozesse repräsentativ beobachtet werden können, nicht aber die tatsächliche Funktionalität des Tiefenlagerbergwerks und seiner Barrieren im Einlagerungsbereich. Damit kann auch kein direkter Beleg für die zumindest anfängliche planungskonforme tatsächliche Anlagenentwicklung erarbeitet werden.

Weitere wissenschaftliche Arbeiten zur Entwicklung von Monitoringsystemen wurden u. a. im international angelegten EU-Verbund-Forschungsprojekt MoDeRn (Monitoring Development for Safe Repository Operation and Staged Closure) bzw. werden derzeit in dessen Nachfolgeprojekt MODERN2020 (Development & Demonstration of Monitoring Strategies and Technologies for Geological Disposal) durchgeführt (vgl. den Beitrag von Jobmann/Liebenstund in diesem Band). Dabei geht es vornehmlich um Monitoringsysteme mit Messgebern, die im Einlagerungsbereich platziert werden, die dann aber nach Versatz und Verschluss des Endlagers nicht mehr zugänglich sind. Längerfristige Funktionstüchtigkeit, Datenübertragung und Energieversorgung sind zentrale Aufgaben. Ein Austausch defekter Messgeber ist nicht möglich. Ebenfalls nicht möglich ist eine Überprüfung der Korrektheit der übermittelten Messwerte.

Im Gegensatz dazu wird im Rahmen des Forschungsprojekts ENTRIA an einem Monitoringkonzept gearbeitet, bei dem eine direkte Überwachung der versetzten Einlagerungssohle mithilfe einer Überwachungssohle und von dort abgeteuften Messbohrlöchern während und auch noch nach Ende der Einlagerungsphase ermöglicht werden soll (Lux et al. 2017a-d). Die Überwachung des Tiefenlagerverhaltens kann bei diesem Monitoringkonzept unmittelbar und großräumig repräsentativ und durch Austausch- oder Ersatzmaßnahmen auch geschützt gegen Messfehler und irreparablen Ausfall der Messsensoren erfolgen. Ein solches direktes Monitoring ermöglicht den Beleg, dass sich das Tiefenlagersystem im Bereich der Einlagerungssohle tatsächlich auch so wie prognostiziert verhält – zumindest über eine später von den dann handelnden Akteuren noch festzulegende, grundsätzlich für die derzeitig vorzunehmenden Systemanalysen aber vorab definierte Beobachtungszeit vor dem endgültigen Verschluss der Entsorgungsanlage. Damit könnten nachfolgenden Generationen

Handlungsräume im Umgang mit den abgelagerten radioaktiven Abfällen eröffnet werden, z. B. die Überführung des Tiefenlagers in ein Endlager mit anschließendem Verschluss bei guter Übereinstimmung der beobachteten Systementwicklung mit der prognostizierten und dem Sicherheitsnachweis zugrunde gelegten Systementwicklung oder eine zügige Rückholung der abgelagerten Abfälle bei messdatenbasiert als kritisch einzustufender Systementwicklung oder bei Entwicklung einer als dann doch besser angesehenen Entsorgungsmöglichkeit. Zu diesem Thema werden mit Blick auf die Zukunftsverantwortung der gegenwärtig lebenden Generationen derzeit noch ethische Diskussionen geführt (Ott/Semper 2017).

Bei Kritikern des im Forschungsprojekt ENTRIA erarbeiteten Monitoringkonzepts besteht die Sorge, dass durch die monitoringbedingte zusätzliche bergbautechnische Infrastruktur für eine direkte Überwachungsmaßnahme die geologische Barriere noch zusätzlich perforiert und dadurch weiter geschwächt wird. Dieser Besorgnis ist durch entsprechende geotechnische und fluiddynamische Analysen zum Systemverhalten einer derartigen zweisöhligen Entsorgungsanlage nachzugehen. Solche geotechnisch und fluiddynamisch fokussierte Analysen können mit dem am seinerzeitigen Lehrstuhl für Deponietechnik und Geomechanik der TU Clausthal entwickelten FTK-Simulator[2] durchgeführt werden (Lux et al. 2015). Hinzuweisen ist in diesem Zusammenhang darauf, dass in einer im Jahr 2008 vom Bundesministerium für Wirtschaft und Technologie (BMWi) herausgegebenen Infobroschüre zur Endlagerung hochradioaktiver Abfälle in Deutschland für das Endlager Gorleben ein zweisöhliges Bergwerk mit Einlagerungssohle und Überfahrungssohle geplant war (BMWi 2008). Dabei war der Überfahrungssohle die Funktion der Erkundung des Salinargebirges im Endlagerbereich und die Funktion einer Abwettersohle zugewiesen worden. Zu diesem Zweck war geplant, Einlagerungssohle und Überfahrungssohle mit großkalibrigen Wetterbohrlöchern zu verbinden. Im Rahmen von ENTRIA ist diese zwischenzeitlich für den Standort Gorleben bezüglich Wettertechnik wieder verworfene Endlagerbergwerkskonzeption in modifizierter Form aufgegriffen und dadurch erweitert worden, dass die konfigurativ schon bestehende Überfahrungssohle in einer weiteren dritten Funktion als Überwachungssohle herangezogen wird.

Die in das Tiefenlager-/Endlagerkonzept implementierte grundsätzlich auch längerfristige Kontrollierbarkeit des Anlagenverhaltens ermöglicht eine Überprüfung der Zuverlässigkeit von Konzeption und technischen Ausführungen auf der Grundlage zuverlässig erfasster Daten. Die dadurch in das Verfahren ein-

2 Der FTK-Simulator (FLAC3D-TOUGH2-Kopplung) besteht aus einer Kopplung der etablierten Simulatoren FLAC3D zur numerischen Modellierung thermomechanischer Prozesse und TOUGH2 zur numerischen Modellierung thermohydraulischer Prozesse.

gebrachte Möglichkeit zur Detektion von Planungs- bzw. Ausführungsfehlern kann aufgrund der zusätzlichen Erkenntnisse zum realen Anlagenverhalten das Vertrauen in die zu treffenden Entscheidungen stärken und die Akzeptabilität eines Endlagerstandortes bei der lokal bzw. regional betroffenen Bevölkerung von vornherein verbessern. Wichtig erscheint in diesem Zusammenhang, dass das jeweilige Monitoringkonzept in seiner technischen Machbarkeit und mit seinen sicherheitstechnischen Folgewirkungen vorgestellt und mit den Beteiligten diskutiert wird, wobei auf seine Möglichkeiten, aber auch seine Grenzen deutlich hingewiesen wird. Darüber hinaus sollte es vorab in seiner Funktionalität belegt werden.

Von den erhobenen Messdaten ausgehend ist von den zukünftig handelnden Akteuren dann allerdings auch zu einem noch festzulegenden Zeitpunkt die finale Entscheidung zu treffen: Rückholung der Abfälle bei nicht plangerechtem Tiefenlagerverhalten oder endgültiger Verbleib der Abfälle bei plangerechtem Tiefenlagerverhalten. Diese mit Blick auf die Möglichkeit zur Demonstration der Anlagensicherheit und die Möglichkeit zur Partizipation an Entscheidungen erwachsenden Vorteile überwiegen bei Weitem den zusätzlichen Aufwand an vorlaufenden generischen Untersuchungen zum Tiefenlager-/Endlagersystemverhalten bei einer auf den ersten Blick grundsätzlich gegenüber den bisherigen konzeptionellen Planungen veränderten Konfiguration der Entsorgungsanlage. Ob zukünftige Generationen von der Möglichkeit eines direkten Monitorings über die Einlagerungsphase hinaus dann tatsächlich auch Gebrauch machen werden und, wenn ja, für welchen Zeitraum, bleibt den Entscheidungen der zukünftigen Generationen überlassen. Die heutige Generation sollte allerdings Handlungsmöglichkeiten zukünftiger Generationen in sicherheitsrelevanten Fragen nicht ohne stichhaltige Begründung (z. B. basierend auf technischen Restriktionen oder auf ethischen Abwägungen zur Verlagerung von Lasten auf zukünftige Generationen) hintenanstellen, sondern ihnen ohne signifikanten Verlust an technischer und ökologischer Sicherheit von vornherein einen möglichst großen Entscheidungsrahmen ermöglichen. Durch eine Standortauswahl ohne Berücksichtigung von Monitoringaspekten könnten Entscheidungsräume schon frühzeitig und später nicht mehr revidierbar reduziert werden. Damit wird deutlich, dass im Rahmen des Standortauswahlverfahrens durch seine prozedurale Ausgestaltung ein Kompromiss zwischen der Gewährleistung bestmöglicher technischer Sicherheit und der Gewährleistung größtmöglicher gesellschaftlicher Gerechtigkeit gefunden werden muss. Gerade bei der generationenübergreifenden Gewährleistung gesellschaftlicher Gerechtigkeit sind vielfältige Aspekte zu bedenken und abzuwägen.

Grundsätzlich wird hier davon ausgegangen, dass durch die Standortauswahl gewährleistet wird, dass sich auch die zweisöhlige Entsorgungsanlage mit Einlagerungs- und Überwachungssohle innerhalb des einschlusswirksamen

Gebirgsbereichs (ewG) (Driftmann 2017) platzieren lässt, wobei die sicherheitstechnischen Vorgaben (BMU 2010) einzuhalten sind. Dieser Ansatz gilt im Grundsatz gleichermaßen für das Tonsteingebirge wie auch für das Salinargebirge. Die aus diesem konfigurativen Ansatz resultierenden Auswirkungen auf den untertägigen Raumbedarf sind, möglichst zeitnah, noch grundlegend und umfassend zu ermitteln. Abbildung 1 zeigt zur Illustration die hier vorgegebene Konfiguration des Tiefenlager-/Endlagersystems mit Überwachungssohle und Messbohrlöchern sowie einem exemplarisch zugeordneten einschlusswirksamen Gebirgsbereich (Lux et al. 2017c).

Abbildung 1: Exemplarische Darstellung des einschlusswirksamen Gebirgsbereichs für ein Tiefenlager-/Endlagerbergwerksystem mit Überwachungssohle

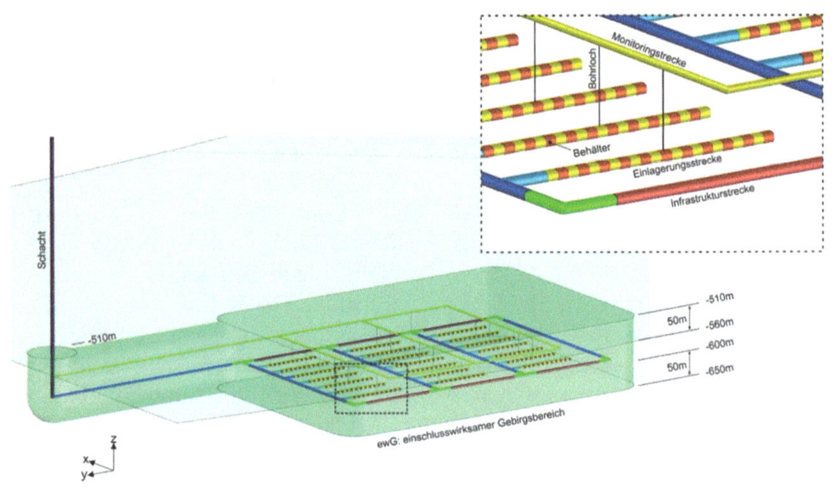

Quelle: Lux et al. 2017c

Die Anordnung einer Überwachungssohle bietet neben dem direkten Monitoring des Tiefenlagerverhaltens im Bereich der versetzten Einlagerungsstrecken und des umgebenden einschlusswirksamen Gebirges durch das Abteufen weiterer Bohrlöcher im Bereich von geotechnischen Barrieren wie z. B. Strecken- und Schachtverschlussbauwerken (s. Abb. 2) zusätzlich die Möglichkeit, innerhalb der Überwachungszeit ihr Verhalten nach dem Einbau im Zusammenwirken mit dem umgebenden Gebirge direkt zu überprüfen und damit die Funktionalität dieser geotechnischen Barrieren zu belegen. Voraussetzung dafür ist, dass geotechnische Barrieren mit zeitnaher Entfaltung ihrer Wirksamkeit im Rahmen von Forschungsvorhaben entwickelt und später auch in die Planung und Ausführung übernommen werden. Eine derartige Konstruktion wird bereits seit einigen Jahren am seinerzeitigen Lehrstuhl für Deponietechnik und Geomechanik der TU Clausthal im Hinblick auf die Funktionalität untersucht (Düsterloh 2014; Lux et al. 2012).

Abbildung 2: Zweisöhliges Tiefenlager-/Endlagersystem mit Anordnung zusätzlicher Bohrlöcher zur Überprüfung der Funktionalität geotechnischer Barrieren, Prinzipdarstellung

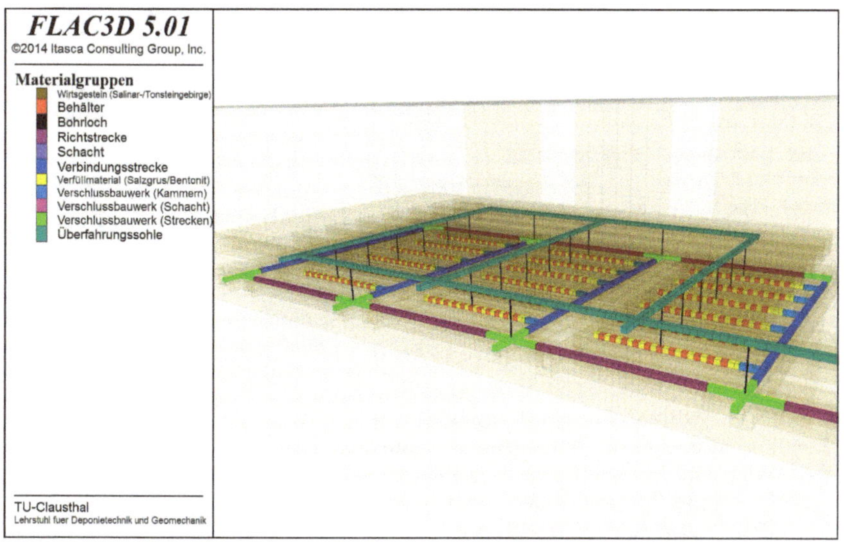

Quelle:Lux et al. 2017c

3 Standortauswahlgesetz und direktes Monitoring – das Yin und Yang eines Versuchs zur Konfliktlösung

Abbildung 3 zeigt zentrale Entwicklungsphasen des Tiefenlager-/Endlagersystems mit direktem Monitoring. Das Monitoring erfolgt begleitend von einer Überwachungssohle und Messbohrlöchern aus während der zeitlichen Entwicklung des Tiefenlagersystems – beginnend mit dem Bau über die Abfalleinlagerung mit zeitnahem Versatz der Einlagerungsstrecken und dem Verschluss der Einlagerungssohle bis zum endgültigen Verschluss des Tiefenlagers aufgrund der Entscheidung, das Tiefenlager in ein Endlager mit der dann nachfolgenden nachsorgefreien Nachverschlussphase zu überführen.

Abbildung 3: Zeitliche Entwicklung eines Tiefenlager-/Endlagersystems mit Berücksichtigung einer langfristigen direkten Monitoringphase vor dem endgültigen Verschluss des Tiefenlagers und der Überführung in ein Endlager

Quelle: Eigene Darstellung, in Anlehnung an Blommaert 2010; Stahlmann et al. 2015

Vor dem Hintergrund der Gefährdungen, die die Funktionalität eines Tiefenlager-/Endlagersystems beeinträchtigen können und die einerseits durch die Schädigung der Barrierenintegrität, andererseits durch die bei Vorhandensein oder Zutritt von fluiden Phasen ermöglichte Mobilisierung und Migration von Radionukliden und auch chemotoxischen Schadstoffen charakterisiert sind, werden in den am seinerzeitigen Lehrstuhl für Deponietechnik und Geomechanik der TU Clausthal bearbeiteten Teilprojekten des Forschungsprojekts ENTRIA neben den thermomechanischen Prozessen insbesondere die fluiddynamischen Prozesse untersucht, die innerhalb von Tiefenlager-/Endlagerbergwerken sowie in der umgebenden Wirtsgesteinsformation auftreten, und zwar sowohl für die Wirtsgesteinsformation Salinargebirge wie auch für die Wirtsgesteinsformation Tonsteingebirge. Diese fluiddynamischen Prozesse in Tiefenlager-/Endlagersystemen im Salinar- bzw. Tonsteingebirge sind sehr komplex aufgrund der sie beeinflussenden thermischen, hydraulischen und mechanischen Einwirkungen sowie der induzierten Wechselwirkungen. Die durchgeführten Untersuchungen leisten damit einen Beitrag zur Verbesserung des Systemverständnisses hinsichtlich des langfristigen Systemverhaltens eines Tiefenlager-/Endlagersystems und damit auch zur Verbesserung der Prognosezuverlässigkeit bezüglich des Verhaltens einer derartigen Entsorgungsanlage für wärmeentwickelnde hochradioaktive Abfälle. Außerdem werden durch die Ermittlung mechanischer, thermischer und hydraulischer Zustandsgrößen Grundlagen für das Monitoring geschaffen. Diese Zustandsgrößen, auf denen auch die radiologischen Sicherheitsanalysen aufbauen, sind als die zentralen Indikatoren der geplanten und genehmigten und damit als sicher konstatierten Tiefenlager-/Endlagerentwicklung anzusehen.

Nicht nur zur Festigung des Vertrauens von Stakeholdern und Zivilgesellschaft in die Zuverlässigkeit der prognostischen Aussagen zur langfristigen Sicherheit der Entsorgungsanlage, sondern auch zur Bestätigung der Entscheidungen der verantwortlichen Institutionen auf Betreiber- und Genehmigungsbehördenseite sich selbst gegenüber erscheint es grundsätzlich geboten, vielleicht sogar erforderlich, soweit sicherheitstechnisch zulässig möglichst direkte Überprüfungsmöglichkeiten zum Verhalten der Entsorgungsanlage während der Einlagerungsphase und auch noch für einige Zeit nach Verschluss der Einlagerungssohle vorzusehen. Eine derartig erweiterte Ausgestaltung war allerdings lange Zeit nicht in das Endlagerkonzept implementiert, bestand doch die feste Überzeugung, durch Anwendung erprobter technischer Verfahren und Erstellung umfangreicher standortbezogener Sicherheitsanalysen eine den sicherheitstechnischen Anforderungen auch langfristig genügende Anlage errichten, betreiben und wieder verschließen und diese Kompetenz ohne weitere Belege in situ überzeugend auch der betroffenen Bevölkerung vermitteln zu können.

Internationalen Tendenzen folgend haben jedoch in den vergangenen Jahren unter dem zentralen Begriff „Fehlerkorrektur" Aspekte wie Reversibilität, Rückholbarkeit und Bergbarkeit sowie Monitoring auch zunehmend Eingang in die Diskussion in Deutschland gefunden (BMU 2010; ESK 2011). Wie dieses Verfahren zur Fehlerkorrektur bei einer Anlagenfehlentwicklung administrativ ablaufen könnte, welche physikalischen Größen als Indikatoren für die Tiefenlagerentwicklung und damit als Entscheidungsgrundlage für eine Fehlererkennung (= Fehlentwicklung der Anlage) herangezogen werden sollten und daher messtechnisch beobachtet werden müssen und welche Bandbreiten in der Entwicklung tolerierbar sind, ist derzeit noch nicht geklärt. Aufgrund des erkannten Forschungsbedarfs ist im Rahmen der Forschungsplattform ENTRIA von vornherein in die zu untersuchenden Entsorgungsoptionen auch die Option „Einlagerung in tiefe geologische Formationen mit Vorkehrungen zur Überwachung und Rückholbarkeit" aufgenommen worden. Die Weitsicht dieses ENTRIA-Ansatzes ist durch die Endlagerkommission, die als bevorzugten Entsorgungsweg das Endlagerbergwerk mit Reversibilität, Rückholbarkeit und Bergbarkeit empfiehlt, nachdrücklich bestätigt worden (BMUB 2017; Kommission „Lagerung hoch radioaktiver Abfallstoffe" 2016).

Die Möglichkeit zur Fehlerkorrektur soll insbesondere deshalb vorgesehen werden, weil angesichts geotektonischer Ungewissheiten und der Temperaturentwicklung eine nur begrenzte Prognosefähigkeit gesehen wird. Ein Anlagenmonitoring wird daher während der Einlagerungsphase und darüber hinaus gefordert.

Um den konzeptionellen Ansatz konfigurativ umzusetzen, ist im Rahmen von ENTRIA als Referenzentsorgungsanlage eine zweisöhlige Bergwerkskonfiguration konzipiert worden (Lux et al. 2017a-d). Das Entsorgungsbergwerk besteht danach aus einer Einlagerungssohle und zusätzlich einer Überfahrungssohle, von der ausgehend Bohrlöcher in die Einlagerungssohle abgeteuft sind. Über diese mit geeigneten Messgebern bestückten Bohrlöcher kann das Verhalten der Einlagerungssohle und ihres Umfeldes auch über die Einlagerungsphase hinaus und damit längerfristig direkt beobachtet werden. Bei vermuteten Fehlmessungen oder dem Ausfall von Messgebern besteht grundsätzlich die Möglichkeit, Messgeber auszutauschen oder zu ersetzen. Die Einlagerungssohle wird in dieser Konzeption wie bislang auch synchron zur Abfalleinlagerung versetzt und verschlossen, während die Überwachungssohle mit den Beobachtungsbohrlöchern längerfristig offen gehalten wird. Damit besteht die Möglichkeit, den verschlossenen Einlagerungsbereich einerseits schon wie bisher frühzeitig in den Zustand einer passiven Sicherheitsgewährleistung zu überführen, ihn andererseits aber auch in seinem nunmehr von außen unbeeinflussten Verhalten längere Zeit direkt zu beobachten und eine den Vorgaben entsprechende Entwicklung explizit festzustellen, nicht nur zu erhoffen bzw. zu vermuten. Bei anforderungsgerechtem Verhalten über einen später noch festzulegenden Beobachtungszeitraum werden dann auch Bohrlöcher und Überwachungssohle sowie die restlichen Schachtbereiche versetzt und verschlossen. Damit wird das zunächst noch überwachte Tiefenlager in das dann nachsorgefreie Endlager transformiert. Folgende Vorteile werden gesehen:

- unmittelbare punktuell/flächenhafte Beobachtung des Tiefenlagerverhaltens,
- zuverlässige Erhebung von entscheidungsrelevanten Messdaten,
- Möglichkeit zum Austausch von defekten Messinstrumenten,
- zeitnahes Erkennen von Fehlentwicklungen des Tiefenlagers,
- messtechnisch belegte Entscheidungsgrundlage für die Endlagerung bzw. Rückholung,
- schrittweises Vorgehen bis zur finalen Implementierung des Endlagers mit durch Feldmessdaten belegter Fehlerkorrekturmöglichkeit,
- keine Notwendigkeit, gleich am Anfang eine irreversible Entscheidung gegenüber zukünftigen Generationen zu treffen,
- größtmögliche Sicherheit und zugleich Flexibilität auch in einem übergenerationell demokratischen Rahmen,
- Erhalt einer Mitwirkungsmöglichkeit zukünftiger Generationen.

Allerdings muss auch angesprochen werden, dass dieser konzeptionelle Ansatz sicherheitstechnisch einige Nachteile aufweist bzw. aufweisen könnte. Zu nennen sind hier:

- größerer untertägiger Raumbedarf in Barrierenqualität für den ewG → Reduzierung der Platzangebote,
- längerfristiger technogener Eingriff in das tiefenlagernahe Gebirge durch Auffahrung und Offenhaltung der Überwachungssohle → höheres betriebliches Risiko,
- zusätzliche Einwirkung auf die geologische Umgebung der Einlagerungssohle innerhalb des ewG/außerhalb des ewG → zusätzliche Barrierenschädigung,
- zusätzliche Versatz- und Verschlussmaßnahmen im Bereich der Messbohrlöcher/Überwachungssohle.

Inwieweit es sich dabei um mehr als nur marginale nachteilige Auswirkungen auf die Standortsuche und die langfristige Sicherheit der Entsorgungsanlage handelt, muss durch entsprechende Untersuchungen ermittelt werden. Erste ausgewählte Befunde werden nachfolgend exemplarisch diskutiert; eine umfassendere Ergebnisdarstellung ist Lux et al. (2017a-d) zu entnehmen.

4 Erste ausgewählte Simulationsergebnisse zur Visualisierung des Tiefenlagerverhaltens

Die Modellierung der in Tiefenlager-/Endlagersystemen ablaufenden geotechnischen und fluiddynamischen Prozesse ist von besonderer Bedeutung bei der Bewertung von unterschiedlichen konzeptionellen Ansätzen, da bei Anwesenheit von fluiden Phasen nach einem Versagen der Abfallbehälter in Verbindung mit einer Mobilisierung von Radionukliden neben einem diffusiv getragenen Radionuklidtransport auch durch Fluidströmungen innerhalb des Tiefenlager-/Endlagerbergwerks sowie im umgebenden Wirtsgestein ein advektiver Transport von Radionukliden bewirkt werden kann. Als Grundlage für die Ermittlung der im Nahfeld des Tiefenlager-/Endlagerbergwerks ablaufenden fluiddynamischen Prozesse ist daher im Rahmen einer Langzeitsicherheitsanalyse die langzeitige Systementwicklung für wahrscheinliche und weniger wahrscheinliche Entwicklungsszenarien zu prognostizieren. Für wahrscheinliche Systementwicklungen (wE) ist in diesem Zusammenhang für den Untersuchungszeitraum von einer Million Jahren der vollständige Einschluss der eingelagerten Abfälle im ewG gefordert (BMU 2010). Für weniger wahrscheinliche Entwicklungen (wwE) wäre der vollständige Einschluss der eingelagerten Abfälle im ewG ebenfalls wünschenswert, gefordert ist allerdings immer noch der sichere Einschluss der eingelagerten Abfälle im ewG, d. h. für den Fall eines Austritts von Schadstoffen (radiotoxisch, chemotoxisch) aus dem ewG in die Biosphäre ist die Einhaltung von vorgegebenen schadstoffbezogenen Grenzwerten nachzuweisen.

Während der Überwachungsmaßnahme und auch während der Nachverschlussphase treten innerhalb des Entsorgungsbergwerks sowie im umgebenden Wirtsgestein neben hier nicht weiter betrachteten chemischen und biologischen Prozessen sehr komplexe physikalische Prozesse auf. Die hier relevanten physikalischen Prozesse sind in mechanische (M), thermische (T) und hydraulische (H) Prozesse zu unterscheiden, die miteinander in enger Wechselwirkung stehen und somit aus Sicht der physikalischen Modellierung als zunächst zweiseitig miteinander gekoppelte Prozesse anzusehen sind. Aus hydraulischer Sicht werden diese Prozesse weiter durch das gleichzeitige Vorhandensein einer Gas- und einer Flüssigphase und damit von zwei Fluidphasen zusätzlich verkompliziert (H2), da die Wechselwirkungen zwischen diesen beiden Phasen ebenfalls noch im Rahmen von rechnerischen Simulationen zum Anlagenverhalten zu berücksichtigen sind. Damit liegen für die Analyse des Systemverhaltens eines Tiefenlager-/Endlagersystems zweiseitig gekoppelte TH2M-Prozesse (= thermisch-hydraulisch-mechanisch gekoppelte Prozesse unter Berücksichtigung von Zweiphasenfluss-Effekten) vor.

Wesentliche Anforderungen an die rechnerischen Simulationen zum Tiefenlagerverhalten sind die Ermittlung der geomechanischen, thermischen und fluiddynamischen Entwicklungen in der versetzten und verschlossenen Einlagerungssohle. Diese Befunde sind Grundlage für den Beleg der langzeitigen Anlagensicherheit und damit der Zuverlässigkeit der konzeptionellen Ausgestaltung der Entsorgungsanlage und auch für den Beleg der bautechnischen Machbarkeit, für die Planung der messtechnischen Ausstattung und für die auf dem Monitoring beruhenden Entscheidungen.

Ein wesentliches Ergebnis der bislang durchgeführten Simulationen ist der Befund, dass für das generische Tiefenlager im Salinar- bzw. Tonsteingebirge bei den jeweilig angesetzten Behälterbeladungen die maximal zu erwartenden Temperaturen in der Überwachungssohle, die hier einen Abstand von 40 m zur Einlagerungssohle aufweist, sowohl in der Wirtsgesteinsformation Salinargebirge wie auch in der Wirtsgesteinsformation Tonsteingebirge im Bereich der festgelegten Grenzwerte (KlimaBergV 1983) bleiben bzw. diese nur in einem geringen, als handhabbar eingeschätzten Maße überschreiten. Möglichkeiten zur Reduzierung der Maximaltemperaturen in der Überwachungssohle sind ein optimiertes Bewetterungskonzept, reduzierte Behälterbeladungen oder ein leicht vergrößerter Abstand zwischen den Behältern innerhalb einer Einlagerungsstrecke bzw. zwischen zwei benachbarten Einlagerungsstrecken, verbunden allerdings mit zusätzlichem untertägigen Raumbedarf.

Ein weiteres wesentliches Ergebnis betrifft den korrosionsbedingten Gasdruckaufbau in der Einlagerungssohle. Abbildung 4 visualisiert exemplarisch für die umfangreichen Simulationsbefunde die langfristig in einem Endlager im Tonsteingebirge zu erwartenden Gasvolumenströme in Einlagerungssohle, Messbohrlöchern und Überwachungssohle. In diesem Fall wird ein Gasgemisch aus Porenluft und Korrosionsgasen aus der Einlagerungssohle durch die in dieser Simulation nur mit Versatzmaterial verfüllten Messbohrlöcher in die ebenfalls mit Versatzmaterial verfüllte Überwachungssohle ausgepresst. Dieses Gasgemisch verteilt sich im Porenraum des Versatzmaterials auf der Überwachungssohle, sodass sich im Vergleich zur Endlagerkonfiguration ohne Überwachungssohle ein weniger intensiver Gasdruckaufbau in der Einlagerungssohle entwickelt.

Abbildung 4: Gasvolumenstrom in einem Tiefenlager-/Endlagerbergwerk im Tonsteingebirge (hier: Einlagerungssohle, Messbohrlöcher und Überwachungssohle) etwa 600 Jahre nach Ende der Monitoringphase

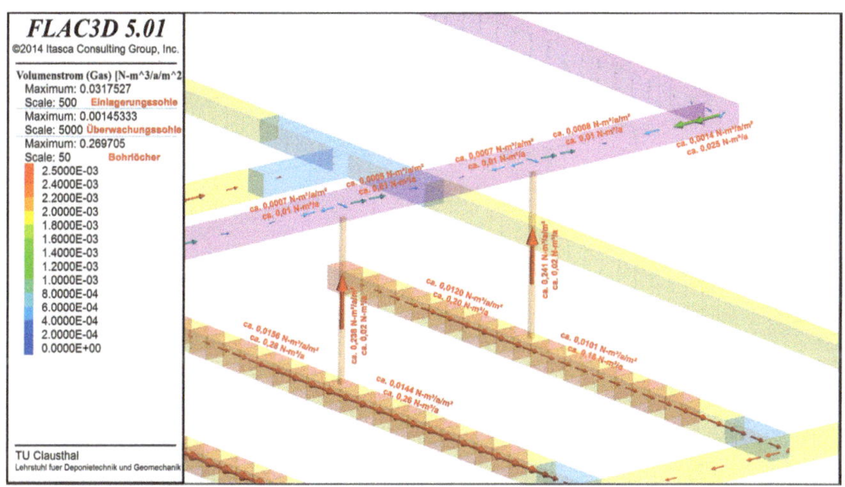

Quelle: Lux et al. 2017d

Ebenso wie für das langzeitige Verhalten des Endlagers werden die die Entwicklung charakterisierenden Zustandsgrößen wie Gebirgs- und Versatzdeformationen, Gebirgs- und Versatztemperaturen, Porenraumdrücke, relative Porenraumfeuchtigkeiten auch für das überwachte Tiefenlagerverhalten ermittelt. Die Entwicklung dieser Zustandsgrößen kann im Tiefenlager dann auch messtechnisch überwacht, mit den planerischen Daten verglichen, von den Akteuren bewertet und den Entscheidungen über das weitere Vorgehen (Endlagerung, Rückholung) zugrunde gelegt werden. Abbildung 5 zeigt exemplarische Zustandsgrößenverläufe.

Abbildung 5: Zeitabhängige Entwicklung von Sättigungsgrad, Porosität, Porendrücken, Temperatur sowie Flüssigkeits- und Gasströmungsraten in ausgewählten Berechnungszonen innerhalb des Tiefenlager-/Endlagerbergwerks

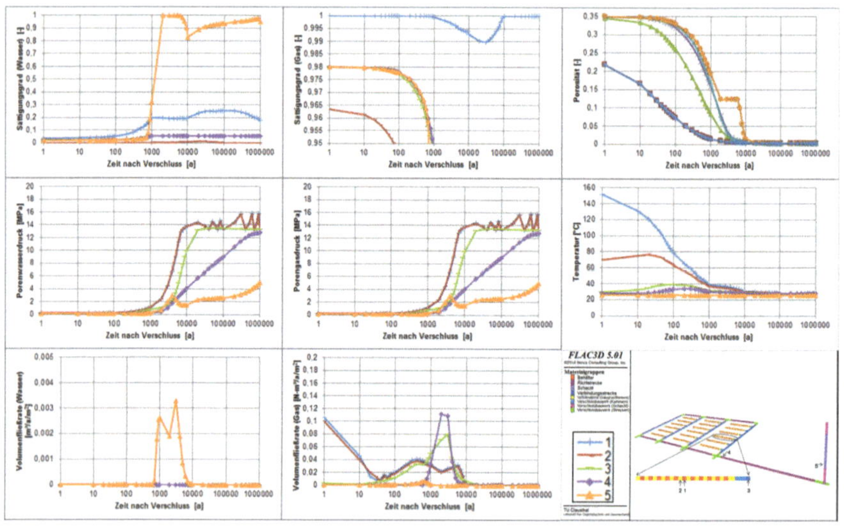

Quelle: Lux et al. 2017b

Weitere umfangreiche Ergebnisse aus den rechnerischen Simulationen für wahrscheinliche und weniger wahrscheinliche Anlagenentwicklungen sind in Lux et al. (2017d) zusammengefasst und dokumentiert.

5 Ausblick auf weitere Arbeiten

Die bisher im Rahmen des ENTRIA-Forschungsprojekts am seinerzeitigen Lehrstuhl für Deponietechnik und Geomechanik der TU Clausthal mithilfe des dort entwickelten FTK-Simulators durchgeführten numerischen Analysen liefern einen wesentlichen Beitrag zur Verbesserung des Prozess- und Systemverständnisses von Tiefenlager-/Endlagersystemen im Salinar- und Tonsteingebirge, insbesondere hinsichtlich der Ermittlung und Visualisierung der im Entsorgungsbergwerk sowie in der umgebenden Wirtsgesteinsformation ablaufenden geotechnischen, thermischen und fluiddynamischen Prozesse. Dabei sind auch erste zentrale Erkenntnisse zu der Fragestellung erarbeitet worden, wie diese Prozesse durch die Anordnung einer zusätzlichen Überwachungssohle oberhalb der Einlagerungssohle in Verbindung mit Messbohrlöchern beeinflusst werden. Hier sind noch weitere umfangreiche Forschungsarbeiten erforderlich, um einerseits den Besorgnissen einer zusätzlichen Barrierenschädigung und der damit verbundenen Implementierung von sicherheitstechnischen Nachteilen zu begegnen und um andererseits die Grundlagen für eine Integration dieses Konzepts mit seinen raumbezogenen Auswirkungen in das Standortauswahlverfahren bereitzustellen.

Das bislang erarbeitete Prozess- und Systemverständnis trägt einerseits zu einer verbesserten Prognosezuverlässigkeit von Tiefenlager-/Endlagersystemen in unterschiedlichen Wirtsgesteinen bei, kann andererseits aber auch in den Dialog mit der von einem Tiefenlager/Endlager betroffenen Bevölkerung eingebracht werden, insbesondere bei der gemeinsam zu führenden Diskussion zur Implementierung von Fehlererkennungs- und Fehlerkorrekturmöglichkeiten in die Anlagenplanung und die damit verbundene Entwicklung einer Monitoringkonzeption. Um diesen Diskurs auf eine hinreichend zuverlässige Grundlage zu stellen, sind weitere numerische Simulationen, aber auch laborative Untersuchungen der als relevant eingeschätzten physikalischen Prozesse durchzuführen. Allerdings kann von der bestehenden Grundlage ausgehend zunächst ein generisch basierter Dialog Technik – Zivilgesellschaft begonnen werden.

Literatur

Blommaert, W. (2010): Reflections on Flexibility, Reversibility, Retrievability by the Belgian nuclear safety authority. Vortrag auf der R&R-Tagung der NEA, Reims, 14.-17. Dez. 2010, FANC, Belgien

BMU – Bundesministerium für Umwelt, Naturschutz und Reaktorsicherheit (2010): Sicherheitsanforderungen an die Endlagerung Wärme entwickelnder radioaktiver Abfälle (Stand: 30.09.2010). https://www.bmuv.de/fileadmin/bmu-import/files/pdfs/allgemein/application/pdf/sicherheitsanforderungen_endlagerung_bf.pdf [Zugriff am 14.02.2022]

BMUB – Bundesministerium für Umwelt, Naturschutz, Bau und Reaktorsicherheit (2017): Gesetz zur Fortentwicklung des Gesetzes zur Suche und Auswahl eines Standortes für ein Endlager für Wärme entwickelnde radioaktive Abfälle und anderer Gesetze (Stand: 05.05.2017). https://www.bgbl.de/xaver/bgbl/start.xav?startbk=Bundesanzeiger_BGBl&start=//*%5b@attr_id=%27bgbl117s1074.pdf%27%5d#__bgbl__%2F%2F*%5B%40attr_id%3D%27bgbl117s1074.pdf%27%5D__1644838320581 [Zugriff am 14.02.2022]

BMWi – Bundesministerium für Wirtschaft und Technologie (2008): Endlagerung hochradioaktiver Abfälle in Deutschland – Das Endlagerprojekt Gorleben. Informationsbroschüre im Rahmen der Öffentlichkeitsarbeit des BMWi, Berlin

Driftmann, C. (2017): Das Endlagerkonzept des einschlusswirksamen Gebirgsbereichs. Eine interdisziplinäre Betrachtung. Braunschweig

Düsterloh, U. (2014): Langzeitsicheres Abdichtungselement aus Salzschnittblöcken – Vorprojekt zur Kalkulation und Qualifizierung der Forschungsarbeiten. Abschlussbericht zum BMWi-Forschungsvorhaben mit dem Förderkennzeichen 02E11223, Clausthal-Zellerfeld

Eckhardt, A. (2017): Monitoring und die geologische Tiefenlagerung in der Schweiz. In: Sammelband Monitoring-Workshop

ESK – Entsorgungskommission/EL – Ausschuss Endlagerung radioaktiver Abfälle (2011): Rückholung/Rückholbarkeit hochradioaktiver Abfälle aus einem Endlager – Diskussionspapier vom 2. September 2011

KlimaBergV – Klima-Bergverordnung (1983): Bergverordnung zum Schutz der Gesundheit gegen Klimaeinwirkungen (Stand: 09.06.1983). http://www.gesetze-im-internet.de/klimabergv [Zugriff am 14.02.2022]

Kommission „Lagerung hoch radioaktiver Abfallstoffe" (2016): Verantwortung für die Zukunft – Ein faires und transparentes Verfahren für die Auswahl eines nationalen Endlagerstandortes. Abschlussbericht; https://www.bundestag.de/resource/blob/434430/bb37b21b8e1e7e049ace5db6b2f949b2/drs_268-data.pdf [Zugriff am 14.02.2022]

Lux, K.-H.; Düsterloh, U.; Dyogtyev, O. (2012): Laborative und numerische Untersuchungen zur Salzgrus-Kompaktion im Verbundsystem Steinsalz-Salzgrus unter THM-Einwirkungen – Orientierende Untersuchungen. Abschlussbericht, Clausthal-Zellerfeld

Lux, K.-H.; Rutenberg, M.; Seeska, R.; Feierabend, J.; Düsterloh, U. (2015): Kopplung der Softwarecodes FLAC^{3D} und TOUGH2 in Verbindung mit in situ-, laborativen und numerischen Untersuchungen zum thermisch-hydraulisch-mechanisch gekoppelten Verhalten von Tongestein unter Endlagerbedingungen. Abschlussbericht zum BMWi-Forschungsprojekt mit dem Förderkennzeichen 02E11041, Clausthal-Zellerfeld

Lux, K.-H.; Wolters, R.; Zhao, J. (2017a): Auf dem langen Weg zu einem Endlager für hochradioaktive Wärme entwickelnde Abfälle – Ein neuer konzeptionell-konfigurativer Ansatz und ein neues Simulationswerkzeug zur Erarbeitung eines verbesserten Prozess- und Systemverständnisses für HAW-Entsorgungsanlagen – ohne und mit direktem längerfristigem Monitoring. Teil I: Endlagerforschung zwischen den Anforderungen nach technischer Sicherheit und sozialer Gerechtigkeit. atw 62(3), S. 185–198

Lux, K.-H.; Wolters, R.; Zhao, J. (2017b): Auf dem langen Weg zu einem Endlager für hochradioaktive Wärme entwickelnde Abfälle – Ein neuer konzeptionell-konfigurativer Ansatz und ein neues Simulationswerkzeug zur Erarbeitung eines verbesserten Prozess- und Systemverständnisses für HAW-Entsorgungsanlagen – ohne und mit direktem längerfristigem Monitoring. Teil II: Der FTK-Simulator als neues Werkzeug zur Analyse fluiddynamischer Prozesse in einer HAW-Entsorgungsanlage. atw 62(4), S. 244–257

Lux, K.-H.; Wolters, R.; Zhao, J. (2017c): Auf dem langen Weg zu einem Endlager für hochradioaktive Wärme entwickelnde Abfälle – Ein neuer konzeptionell-konfigurativer Ansatz und ein neues Simulationswerkzeug zur Erarbeitung eines verbesserten Prozess- und Systemverständnisses für HAW-Entsorgungsanlagen – ohne und mit direktem längerfristigem Monitoring. Teil III: Ein neues konzeptionelles und konfiguratives Konzept zur Entsorgung hochradioaktiver Wärme entwickelnder Abfälle. atw 62(5), S. 317–326

Lux, K.-H.; Wolters, R.; Zhao, J.; Rutenberg, M.; Feierabend, J.; Pan, T. (2017d): TH2M-basierte multiphysikalische Modellierung und Simulation von Referenz-Endlagersystemen im Salinar- und Tonsteingebirge ohne bzw. mit Implementierung einer Möglichkeit des längerfristigen Systemverhaltens auch noch nach Verschluss der Einlagerungssohle. Ein Beitrag zur Verbesserung der Robustheit von Sicherheitsfunktionen mit sehr hoher Relevanz im Hinblick auf die Entwicklung von Bewertungsgrundlagen zum Vergleich von Entsorgungsoptionen. ENTRIA-Arbeitsbericht-07, Clausthal-Zellerfeld

Nagra – Nationale Genossenschaft für die Lagerung radioaktiver Abfälle (2014): Modelling of Radionuclide Transport along the Underground Access Structures of Deep Geological Repositories. NTB 14–10, Wettingen

Ott, K.; Semper, F. (2017): Nicht von meiner Welt. Zukunftsverantwortung bei der Endlagerung von radioaktiven Reststoffen. GAiA 26(2), S. 100–102

Stahlmann, J.; Leon-Vargas, R.; Mintzlaff, V. (2015): Generische Tiefenlagermodelle mit Option zur Rückholung der radioaktiven Reststoffe: Geologische und Geotechnische Aspekte für die Auslegung. ENTRIA-Arbeitsbericht-03, Braunschweig

Volker Mintzlaff, Rocio Paola León Vargas, Joachim Stahlmann, Ida Epkenhans

Requirements for geotechnical monitoring of deep geological repositories with retrievability

1 Motivation for a geotechnical monitoring[1] program in a repository

What should be the most important reason to retrieve emplaced high-level radioactive waste (HAW) from a geological repository? No one needs this waste back![2] The main reason for retrieval is the possibility to react in case of unexpected developments that could question the long-term safety of the repository system. To ensure retrievability, it is necessary to have information about the state of the repository system. Monitoring is the key to gathering information about the repository. The collected data can help verify models, maintain and establish confidence in the functionality of the repository system, and is essential to provide a sound basis for deciding on whether the repository can be closed or whether waste retrieval must be initiated. By means of near-field monitoring, it is possible to get information about the state of geotechnical and geological barriers.

The possibility of a safe retrieval of the emplaced waste adds flexibility to the disposal program and thus gives freedom of decision. Hence, a deep geological repository with retrievability compromises the (passive) long-term safety of final disposal and the flexibility to act actively as in long-term surface storage.

2 The lifespan of a repository with retrievability

The lifespan of a repository with retrievability covers three main phases: pre-operational phase, operational phase, and post-operational phase (see Figure 1). In the pre-operational phase, the site selection process and the licensing procedure including safety assessments will be performed. In the operational

1 In this paper, "geotechnical monitoring" is defined as technical provisions for systematic measurement of the in-situ state of the mine of a high-level radioactive waste repository.

2 Possible reuse or future technologies such as partitioning and transmutation (P&T) to treat the waste are not considered in this contribution.

phase, the repository mine will be constructed, the waste emplaced, and the emplacement drifts backfilled. After emplacement of the waste, an optional monitoring phase can be implemented. At the end of the operational phase, all monitoring boreholes, drifts, and shafts will be backfilled; the repository will be converted to a final disposal repository; the mine will be closed. In the post-operational phase, the HAW could only be recovered by building a new mine.

Figure 1: Lifespan of a repository with retrievability

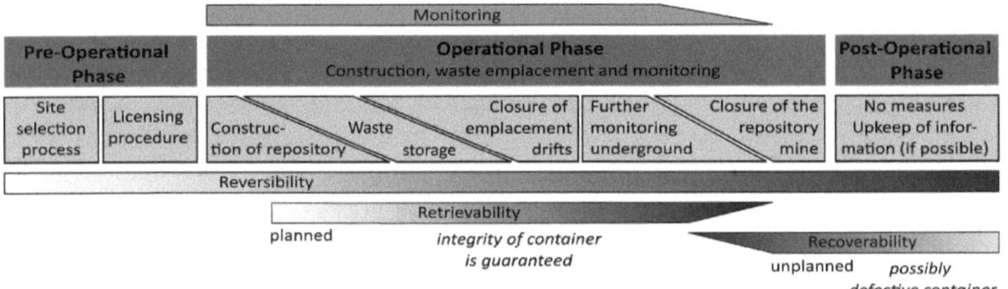

Source: based on Stahlmann et al. 2015

Since monitoring is a complex matter, it needs to be planned well in advance, which means that the preparations for the monitoring need to be adapted already at the site selection stage. The implementer, the regulator, and the stakeholders should be integrated at this point in the project.

Monitoring data is the only source of information about the state of the repository mine. Sensor or data transmission failures could be critical. From a technical point of view – if there is no or uninterpretable information about the state of the mine – a "worst case" cannot be identified and therefore not disclosed. Consequently, the emplaced waste must be retrieved, endangering the staff who have to retrieve the waste. This implies that robust sensors and robust data transmission systems must be developed before repository construction and underlines the importance of careful planning.

Geotechnical monitoring of a repository with retrievability must be implemented from the beginning of the operational phase in order to collect reference values (see Eckhardt in this volume). These values are essential for the interpretation of the collected data during the entire period that the repository is open, as developments can only be detected by comparison of presently measured values with previously measured ones. This limits (but does not exclude) the possibility of changing the monitoring program once the repository has been constructed. Monitoring should continue during the operational and post-operational phases until the repository is closed. The collected data from the monitoring program will serve as a sound basis for a geotechnically-based decision-making on retrieval/closure. Not the subject of this article, but nevertheless important, is social monitoring of the disposal program, which should be considered during the entire lifespan.

3 Generic repository designs to evaluate the impact of retrievability and monitoring

To evaluate monitoring concepts for different host rocks, a generic repository design[3] was developed in the ENTRIA project. It consists of a two-level mine: an emplacement level, where the waste is stored in self-shielding casks in horizontal drifts, and a monitoring level situated above, where the near-field measurements take place (see Figure 2). To reduce the impact of further excavation on the host rock, the amount of monitoring drifts must be minimized. Therefore, each monitoring drift is used to monitor two emplacement drifts. The emplacement drifts will be backfilled immediately after emplacement. To ensure the proper functionality of the backfill and to reduce deviatoric stress in the host rock, the emplacement drifts will be closed with a sealing structure. The underground infrastructure, consisting of the monitoring level and the shaft and the infrastructure for a potential retrieval at the emplacement level, will be kept open to facilitate retrieval. The geometrical details of the models are presented in Stahlmann et al. (2015).

Deep geological repositories with retrievability must meet two objectives: The first objective, as for a final disposal repository, is long-term safety. The second objective is to ensure the accessibility of the HAW and, thus, continued operational safety during the period the repository is open. A monitoring program is needed to gather information on the state of the repository. The need to keep the mine open for a longer period of time compared to the final disposal concept results in a trade-off: Open drifts degrade the geological and geotechnical barriers. Furthermore, the time frame of stability of open drifts also limits retrievability. Therefore, it must be ensured that there is minimal loss of integrity of these barriers during the operational and post-operational phases. To integrate these objectives into the generic repository design, two main functions were implemented:

---

3 This generic repository is not adapted to a certain site with its specific geology and country-specific regulations. Therefore, many influences on the design can be neglected, giving a conceptual view of retrievability.

(1) The emplacement drifts are backfilled immediately after waste emplacement and closed with a sealing structure, consisting of an abutment and a seal. The abutment reduces stress rearrangement in the emplacement drifts in order to achieve a quick and safe enclosure of the HAW in the host rock. Furthermore, the backfill minimizes the underground openings in the mine and provides radiation protection and protection from the heat generated by the emplaced waste.
(2) The monitoring drifts are kept open to collect data on the development of the open mine and the emplacement drifts through boreholes. As part of the closure of the repository, all cabled monitoring equipment and monitoring boreholes will be backfilled.

Figure 2: Generic model for a deep geological repository with retrievability

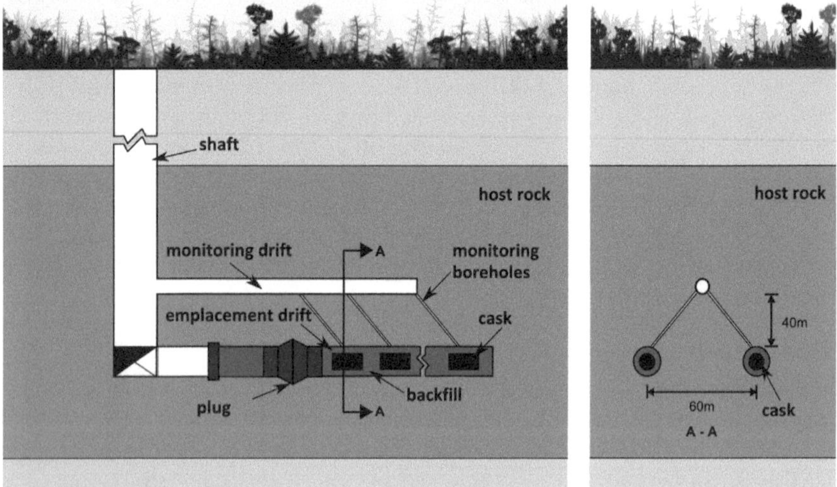

Source: Stahlmann et al. 2016

In general, the generic repository models can be fitted to every host rock under discussion in the site selection process in Germany. To perform a comparison between these host rocks, the backfill, the abutments, and the seals have to be adapted to the host rocks. For instance, in the case of salt, the backfill material is crushed salt and the abutment and seal are made of salt cement; no support of the drifts is needed. The other host rocks (clay, shale, and crystalline hard rock) have a bentonite-based backfill and seal; the abutment must be made of low-pH concrete. In clay and shale rock, drift supports, such as low-pH shotcrete supports, are needed.

The properties of the different host rocks determine the development of the respective repository system. Every host rock has its unique characteristics. From the geotechnical point of view, these characteristics can be described with the following parameters:

- primary state of stress
- plasticity and creep capability
- joint system

These determine the long-term self-support of the host rock, the convergences in the drifts, the extension and growth of the excavation damage zone (EDZ), and the amount of groundwater and its flow in the repository system. Heat transfer in the repository system is determined by thermal conductivity and capacity of the host rock.

A very important influencing factor in the repository is the heat generated by the emplaced waste. The heat affects the mechanical properties of the rock and the supporting system (if needed) in the repository mine. Furthermore, it limits the feasibility of work in the drifts if the air temperature exceeds certain limits.

Regarding the objective of long-term accessibility of the waste, good long-term self-support and low convergences are the main geotechnical aspects with the most desirable, positive influences. Crystalline hard rock is the most suitable host rock for this part. Due to its capacity to support itself without additional technical measures, convergences during the operating period can be neglected. Shale has a moderate self-supporting capacity and low plasticity. Clay, on the contrary, has a very low self-supporting capacity, thus requiring an extensive support system. Salt can support itself only for a limited time due to the convergences based on its creeping capability. Therefore, in rock salt, the period of opening is limited.

The second objective of long-term safety is determined by the amount of groundwater and the groundwater flow in the repository system. Groundwater is the primary transport path for radionuclides. In highly impermeable rocks, the fracture network determines whether the groundwater can flow or stagnates. Open fractures are usually found in crystalline hard rock and in shale. In crystalline hard rock, these fractures cannot seal themselves and remain open. In clay, there is a latent fracture network. Due to its plasticity, fractures in clay can self-seal, as they do to a lesser extent in shale. Rock salt has no fracture network due to its creeping capability. Excavation cracks can self-seal if the initial state of stress is restored. In rock salt deposits, there is no groundwater except small inclusions. Groundwater issues must be considered and evaluated for a repository system in clay and shale (stagnating groundwater, possible groundwater flow in fissures in shale) as well as in the case of large amounts of groundwater in crystalline hard rock.

4 Setting up a monitoring program

Considering the main factors influencing repository safety, a monitoring program must answer the following questions: why, where, when, what, and how to measure (according to AITEMIN 2011). These questions were set up in the European MoDeRn project, which is described in the contribution of Jobmann and Liebenstund in this volume.

Why to measure?

The scientific aim is to verify the repository models and to confirm or change the model expectations we made before the implementation of the repository. Monitoring is the only way to get the information that models may not apply. Because of this aim, it is important to measure everywhere reasonably possible in the repository system to learn whether the safety functions of the repository are harmed. This aim also includes demonstrating that the repository system functions as approved. Faults in the installation of geological and engineered barriers should be detected, too.

Where and when to measure?

The infrastructure drifts and their surroundings can be instrumented directly. In the closed emplacement drifts, measurements are possible from boreholes drilled from the monitoring drift at the monitoring level. To reduce perforation of the host rock, two emplacement drifts are instrumented by a monitoring drift. If the geology is homogenous – which is not often the case – it is possible to reduce instrumentation to one borehole per homogenous area. Measurements should be implemented from the beginning of the construction of the repository to collect reference values and should be maintained until closure. Monitoring is seen as a process: Starting with the construction of the repository in the operational phase, the data collected can show whether conceptual adjustments are necessary for the safe operation of the repository. In the next steps, during waste emplacement and the post-operational phase, the monitoring data provides a basis for deciding on whether the repository is developing as expected or whether the waste may have to be retrieved. Also, it can help determine how long the repository can be kept open in the operational phase after waste emplacement. Due to the long timespan, the reliability of the sensors and data nodes is a key point. There are, so far, no sensors with an expected lifespan covering the entire monitoring phase. The instrumentation must be repairable and replaceable. Therefore, boreholes reaching the sensors are inevitable. Data transfer and power supply can be carried out in the same way, which otherwise would pose new technical problems (related to final disposal monitoring, see Jobmann et al. 2011 and Jobmann/Liebenstund in this volume).

What to measure?

This has to be decided for each host rock separately. For example, in crystalline host rocks, groundwater flow and water pressure are very relevant; in rock salt, groundwater would not need to be measured. Relevant parameters for a geotechnical assessment of the state of the repository are: strain, stress, and convergence. It would be convenient to measure changes in the EDZ as well as in the permeability of the host rock and the geotechnical barriers; however, this is only possible indirectly. Regarding the robustness of the monitoring system, the most simple and reliable parameters should be instrumented, such as displacements in the host rock and pore pressures. Temperature measurements are very feasible and should therefore be performed in each host rock as well.

Figure 3 gives an overview of the development of the relevant parameters during the construction of the repository in three steps: In the first step, the shaft and the monitoring drift will be constructed. The convergences of these constructed voids will be measured. The state of stress in the host rock will be measured, which is expected to be slightly different from the primary state of stress. The temperature change in the rock, the air humidity, and, if feasible, the alteration in permeability should be instrumented. In groundwater-bearing rocks, the groundwater inflow should be measured. In the second step, the emplacement level will be constructed. Convergences and alterations in the permeability of the EDZ must also be measured in the new drifts. Measurements will continue during the complete operational phase. In the third step, measurements must also be made in the backfill of the emplacement drifts to get a complete picture of the development of the repository system. In rock salt, the development of compaction of the crushed salt is relevant. It can be measured indirectly by seismic methods. In the other host rocks, the swelling pressure of bentonite as backfill material can facilitate the clogging of voids. Rock stresses and their development around the emplacement and infrastructure drifts can serve as an indication of where the rock is overstressed.

The development of convergences is relevant in rock salt. If there is only a simple wall holding the backfill in place, it could be displaced. These displacements are caused by the rock converging into the backfill and pushing it toward the remaining void behind the wall. The wall could move into the void and eventually break. This would result in degradation of the host rock salt and, consequently, in higher permeability of the EDZ. To avoid this process, an abutment is needed: It keeps the backfill in place so that the backfill can generate a back pressure to the rock stress over time.

Figure 3: The three construction phases of the generic repository model

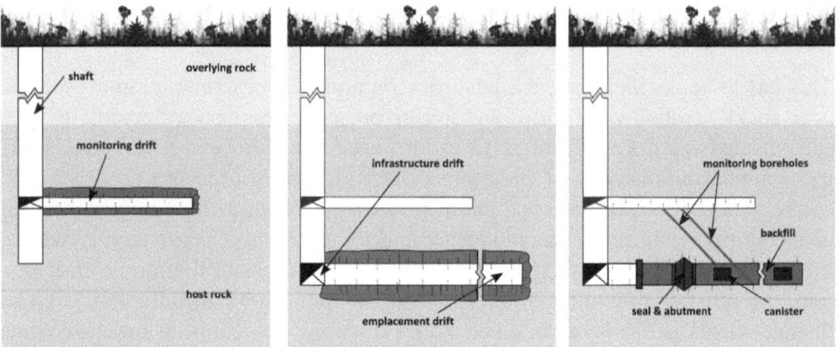

Source: Own illustration

In other soft rocks, such as clay and shale, bentonite is used as backfill material. When coming into contact with water, it will swell and seal remaining voids. It will also build up a swelling pressure. If the backfilled drift is closed only by a wall, the wall will be displaced and the swelling pressure will decrease. Due to the loss of the swelling pressure, the convergences into the backfilled void will not be reduced, the EDZ will increase, and the host rock barrier will deteriorate. To avoid these processes, an abutment is required as in rock salt. The swelling pressure will be able to build up and lead to a back pressure to the rock stress, reducing the convergences (Stahlmann et al. 2016). In crystalline host rocks, an abutment is required to ensure sealing of the fissures in the host rock.

5 Conclusion

Monitoring is a very important issue, as it enables the implementer and the regulator to get information about the state of geotechnical and geological barriers of the repository. The information collected provides a sound basis for deciding on whether the repository is ready for closure or retrieval. It can be useful for validating and/or adopting the scientific models and for communicating with interested stakeholders as well.

The generic repository design presented offers the possibility to evaluate the main parameters in every host rock. Based on this conceptual framework, it is possible to identify positive and negative aspects of retrievability and assess their impact on operational and long-term safety. This information can be used in early stages of a site selection process. In terms of host rocks, crystalline hard rocks are more suitable for retrievability compared to clay, shale, and rock salt. However, this comes at the expense of long-term safety due to the lower barrier function of the rock itself. In principle, retrievability is possible in all the other host rock systems, with impacts on the long-term safety of the disposal.

There are further relevant aspects of geotechnical monitoring from a societal point of view. It is about organization: Who will be in charge of the monitoring and who will control the monitoring staff? How will the data be interpreted? Will politicians and other decision-makers accept the interpretation? To what extent will the interpretation be communicated – will there be public access only to the interpreted data or to the raw data too? Will there be sufficient interest from society to continue monitoring over the entire lifespan of the repository?

Monitoring of deep geological repositories is therefore very complex, both technically and socially. Monitoring provides the possibility to intervene when the repository does not develop as expected. Compromises have to be made between short-term monitoring evaluation and the option to intervene, on the one hand, and reduced long-term safety, on the other hand. Another compromise that must be considered is how close to the geotechnical barriers to measure. Measuring too close can cause further perforation of these barriers, affecting long-term safety.

Monitoring and retrievability increase the technical risk of disposal because more open drifts are needed and these drifts remain open longer. This is the price to pay for the ability to react immediately in the event of an unexpected development. The analysis of the generic repository design has shown that this increase in risk is technically manageable. Society has to decide whether the value of the gains in confidence and ability to act in the repository system is high enough to partially compromise the – not directly verifiable – passive safety of a repository.

References

AITEMIN (2011): MoDeRn Technical Requirements Report. Deliverable D2.1.1., 03.01.2011

Jobmann, M.; Eilers, G.; Haverkamp, B. (2011): Überwachung eines Endlagers für hochradioaktive Abfälle in Deutschland – Möglichkeiten und Grenzen. Internationale Zeitschrift für Kernenergie (ATW) 56(11), pp. 629–635

Stahlmann, J.; Leon Vargas, R.; Mintzlaff, V. (2016): Geotechnische und geologische Aspekte für Tiefenlagerkonzepte mit der Option der Rückholung der radioaktiven Reststoffe. Bautechnik 93(3), pp. 141–150

Stahlmann, J.; Mintzlaff, V.; Leon Vargas, R. (2015): Generische Tiefenlagermodelle mit Option zur Rückholung der radioaktiven Reststoffe: Geologische und Geotechnische Aspekte für die Auslegung. ENTRIA-Arbeitsbericht-03, Braunschweig

Thomas Hassel, Ansgar Köhler

Die Bedeutung der Rückholbarkeit als Option im Zusammenhang mit dem Wunsch und der Notwendigkeit eines Langzeit-Monitorings

1 Einleitung

Bei der Planung eines geologischen Tiefenlagers als Endlager für hochradioaktive Abfälle hat die Forderung nach einer Interventionsmöglichkeit für den Fall einer nicht erwartungsgemäßen Entwicklung des Lagersystems in den letzten Jahren erheblich an Bedeutung gewonnen. Das BMU hat bereits 2010 vorgeschrieben, dass eine Handhabbarkeit der Abfallbehälter über einen Zeitraum von 500 Jahren gegeben sein muss; diese Forderung wurde im Standortauswahlgesetz bestätigt (BMU 2010; StandAG 2017). Gerade in den ersten Jahrzehnten nach dem Verschluss der Einlagerungsstrecken ist die vollständige hydraulische Wirksamkeit der errichteten geotechnischen Barrieren noch nicht gegeben (Brasser/Miehe 2008; Müller-Lyda 1999). Die Sicherheit und Funktionsfähigkeit des Tiefenlagers ist zu diesem Zeitpunkt in großem Maße von der Integrität der technischen Barrieren, also dem System aus Einlagerungsbehälter und direkt umgebenden Verfüllmaterial, abhängig.

Die Option der Rückholung setzt eine sichere Handhabbarkeit und somit auch die strukturelle Unversehrtheit der Behälter voraus. Die Beobachtung der technischen Barriere besitzt demnach bei einem Tiefenlager mit der Option der Rückholbarkeit eine besonders hohe Relevanz. Im Rahmen dieses Beitrags werden die Randbedingungen und Herausforderungen, die sich für das Monitoring an der technischen Barriere ergeben, beschrieben.

Der Begriff Monitoring beschreibt im Allgemeinen die systematische Erfassung, Beobachtung und Überwachung von Vorgängen oder Prozessen mithilfe von Beobachtungssystemen. Ziel ist es, Daten über ein System zu gewinnen, anhand derer die Entwicklung desselben beobachtet werden kann, sodass ein rechtzeitiger Eingriff möglich wird.

Bei der Auswahl geeigneter Messgrößen ergibt sich oft ein Zielkonflikt zwischen der Relevanz der ausgewählten Messgröße und ihrer technischen Erfassbarkeit im gegebenen Umfeld. Zusätzlich muss eine Interventions- bzw. Eingriffsmöglichkeit gegeben sein, falls eine Abweichung von den Sollparametern erkannt wird. Am Beispiel von zwei Eigenschaften der technischen

Barriere, der Dichtheit der Behälter und der Temperaturentwicklung, soll dies dargestellt werden.

Der dichte Einschluss der radioaktiven Abfälle stellt die Kernanforderung an den Behälter dar. Ist dieser nicht mehr gewährleistet, kann ein Austrag von Radionukliden aus dem Behälter erfolgen und eine sichere Handhabung ist dann nicht mehr möglich. Hieraus resultiert die Forderung, dass Prozesse, die zu einer Schädigung der Barrierefunktion des Behälters während der Rückholphase führen, frühzeitig erkannt werden müssen, sodass eine Intervention noch möglich ist. Die Behälter sind nach ihrer Einlagerung nicht mehr zugänglich, dadurch wird es wesentlich schwieriger, Informationen über ihren Zustand zu gewinnen. Der Verlust der Einschlussfunktion des Behälters lässt sich beispielsweise mittels eines Druckschalters zur Überwachung des Innendrucks im Behälter erfassen. Das Konzept wird bereits seit Jahren zur Überwachung von Castor-Behältern in Zwischenlagern eingesetzt (Hofmeister 2015). Die Messgröße „Behälterinnendruck“ ist für das Behältermonitoring im Endlager aber nicht geeignet, da der Integritätsverlust des Behälters anhand dieser Messgröße erst dann erkannt werden kann, wenn der Behälter bereits undicht geworden ist. Eine Intervention ist zu diesem Zeitpunkt nicht mehr gefahrlos möglich, da infolge mechanischer Belastungen bei einer Rückholung mit einer Freisetzung von Radionukliden gerechnet werden muss. Auch die kabellose Übertragung des Sensorsignals aus dem Behälter ist nicht geklärt.

Die Prozesse, die zu einem Verlust der Behälterintegrität führen, können somit im Tiefenlager nur anhand der Beobachtung von Einflussgrößen und durch die Überwachung resultierender Wechselwirkungen mit dem Behälterumfeld erkannt werden. Beispielsweise ist anhand der Entwicklung der Feuchtigkeitssättigung im direkten Behälterumfeld ein Rückschluss auf die korrosiven Vorgänge an den Behältern möglich. Auch durch die Erfassung von Korrosionsprodukten wie z. B. Wasserstoff können Anhaltspunkte über den Zustand der Behälter gewonnen werden. Die Herausforderung besteht in der Identifikation und quantitativen Erfassung der relevanten Einfluss- und Messgrößen. Weiterhin müssen geeignete Sensorsysteme zur Erfassung der Messwerte zur Verfügung stehen. Wichtigster Punkt dabei ist die Entwicklung von Konzepten zur Interpretation der Messdaten. Erst hierdurch wird ein Vergleich der tatsächlichen Entwicklung der technischen Barriere mit der beispielsweise im Rahmen eines Langzeitsicherheitsnachweises prognostizierten Entwicklung für den Entscheider ermöglicht.

Anders gestaltet sich die Situation bei der Überwachung der Temperaturentwicklung im Endlager. Der Behälter stellt eine Wärmequelle dar, die das direkte Umfeld erheblich erwärmt. Im Falle der Rückholung muss zuerst das Temperaturniveau im Bereich der Behälter aus Gründen des Arbeitsschutzes abgesenkt werden, bevor mit dem Abtransport der Behälter begonnen werden kann. Das Temperaturniveau stellt somit einen wichtigen Einflussfaktor für die Planung der Rückholung und eine der zentralen Messgrößen des Behältermonitorings dar. Die Temperaturgrenzen im Endlagersystem werden von der maximalen Temperaturfestigkeit der Materialien der umgebenden geologischen sowie geotechnischen Barriere, welche beispielsweise für das Verfüllmaterial Bentonit bei maximal 100 °C liegt, definiert (Stahlmann et al. 2015). Die Abfallgebinde besitzen eine wesentlich höhere Temperaturfestigkeit. Diese wird bei den meisten Behälterkonzepten durch den Schmelzpunkt des als Abschirmmaterial eingesetzten Polyethylens begrenzt, die bei etwa 160 °C liegt (Voßnacke et al. 2004). Die Grenzwerte für die Temperatur orientieren sich an dem Grenzwert der geotechnischen Barriere. Falls dieser überschritten wird, ist noch nicht mit einer Schädigung der Abfallgebinde zu rechnen und eine Intervention beziehungsweise Rückholung der Behälter ist weiterhin möglich. Die Temperaturentwicklung ist somit ein wichtiger Parameter für die Rückholung der Behälter. Sie ermöglicht wiederum aber keinen Rückschluss auf die Behälterintegrität.

Weitere wichtige Parameter zur Beschreibung des Zustands der technischen Barriere und für die Planung einer Rückholung stellen z. B. die Lage des Behälters im Endlagersystem sowie die einwirkenden gebirgsmechanischen Kräfte dar. Auch für diese Parameter sind im Rahmen der Entwicklung eines Monitoringkonzepts geeignete Messgrößen zu bestimmen.

2 Monitoring der technischen Barriere im Umfeld eines geologischen Tiefenlagers

In der Industrie, z. B. im Fahrzeugbau sowie im Bauwesen, stellt Monitoring heutzutage eine häufig angewendete Maßnahme dar, um den Zustand komplexer Bauteile oder Strukturen zu erfassen und so Aussagen über die voraussichtliche weitere Entwicklung dieser Komponenten zu treffen. Die Anforderungen, die sich aus dem geologischen Umfeld und durch die Merkmale der geotechnischen und technischen Barriere an die Komponenten des Monitoringsystems ergeben, unterscheiden sich stark von denen des Anlagen- oder Bauwerkmonitorings. Somit können die in diesen klassischen Anwendungsbereichen des Monitorings gesammelten Erfahrungen nur zum Teil auf Monitoringkonzepte im Umfeld eines Endlagers übertragen werden.

Zur Datenerfassung werden beim Monitoring Messsysteme verwendet, die jeweils aus einem oder mehreren Sensorsystemen bestehen. Ein Sensorsystem umfasst dabei einen oder mehrere Einzelsensoren zur Erfassung der Messgröße sowie Komponenten zur Energieversorgung, Datenübertragung und gegebenenfalls Datenverarbeitung. Die Kernanforderungen an die Messsysteme definieren sich über die zuvor festgelegten Messaufgaben sowie das Umfeld der Messung. Die Sensorsysteme im Tiefenlager müssen die zu erfassenden Daten über mehrere Jahrzehnte zuverlässig und hinreichend genau aufnehmen und fehlerfrei an das Messsystem übermitteln. Die Sensoren müssen für den Einsatz am geplanten Einbauort geeignet sein und unter den gegebenen Umgebungsbedingungen langfristig zuverlässige Messdaten ermitteln.

In den meisten klassischen Einsatzfeldern des Monitorings, wie beispielsweise beim Monitoring von Industrieanlagen, sind die zu überwachenden Bauteile gut zugänglich. Daher besteht die Möglichkeit, die Sensoren unter Betriebsbedingungen zu überprüfen oder zu kalibrieren und defekte Sensoren einfach auszutauschen. Auch die Übertragung der Sensordaten an das Messsystem stellt hier zumeist keine große Herausforderung dar, da entweder eine direkte Verdrahtung des Messwertaufnehmers mit dem Messsystem möglich ist oder bei einer Datenübertragung per Funkverbindung eine zuverlässige Energieversorgung für den Sensor sichergestellt werden kann.

Für das Monitoring in einem Tiefenlager gelten komplexere Randbedingungen. Bei der geologischen Tiefenlagerung wird der Ansatz gestaffelter Barrieren („defense in depth“) verfolgt. Wie in Abbildung 1 ersichtlich ist, existieren im Tiefenlager eine Reihe unterschiedlicher Barrieren. Diese können ihre planmäßige Funktion nur dann erfüllen, wenn sie homogen aufgebaut sind und keine unnötigen Durchörterungen (z. B. Bohrungen für Kabel) aufweisen. Das innerste Barrieresystem ist der Endlagerbehälter. Für die Überwachung dieser technischen Barriere ergeben sich aus der Forderung der Unversehrtheit der umgebenden Barrieren hohe Anforderungen an das Monitoringsystem.

Die Behälter werden in die Einlagerungsstrecke eingebracht, die direkt im Anschluss verfüllt und verschlossen wird. Ab diesem Zeitpunkt sind die Behälter nicht mehr zugänglich. Dies bedeutet, dass an allen Sensorsystemen, die in räumlicher Nähe der Behälter positioniert sind, keine Wartung mehr durchgeführt werden kann. Auch ein Austausch ist nicht mehr möglich. Die erste Kernforderung an die Sensorsysteme für das Behältermonitoring ist demnach die Forderung nach einer sehr hohen Robustheit und einer langfristig hohen Zuverlässigkeit aller Komponenten.

Auch die weiteren Anforderungen an das Monitoringsystem für die technische Barriere ergeben sich aus dem Verbot der Perforation der geologischen sowie geotechnischen Barrieren (Bohner 2014; Solente et al. 2013). Eine Verkabelung von Sensoren durch die Barrieren hindurch ist nicht möglich, sodass eine autarke Energieversorgung sowie eine kabellose Datenübertragung für die Sensorsysteme erforderlich sind.

Abbildung 1: Schema des gesamten Barrieresystems im geologischen Tiefenlager mit idealisiertem Endlagerbehälter

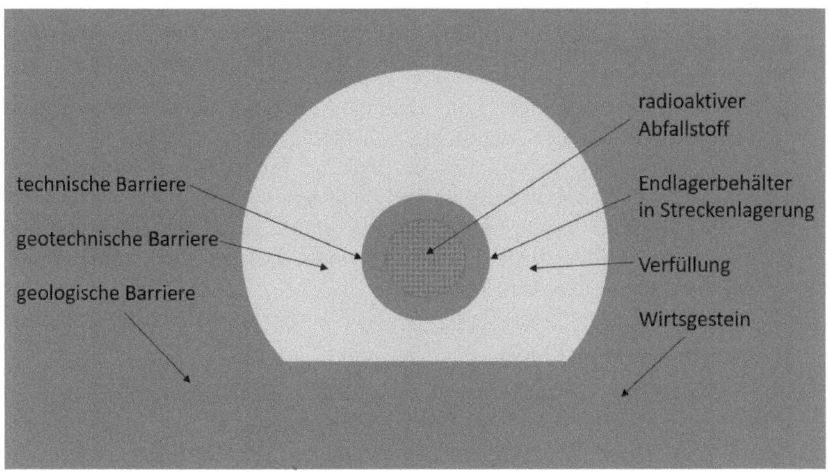

Quelle: Eigene Darstellung: Institut für Werkstoffkunde; Leibniz Universität Hannover

Die drei Anforderungen

- Zuverlässigkeit des Sensorsystems,
- autarke Energieversorgung,
- kabellose Datenübertragung

stellen durch den langen Zeitraum, über den die Beobachtung erfolgen muss, eine große Herausforderung dar. Die Betriebsphase des Lagers, von der Einlagerung des ersten Behälters bis zum Verschluss des letzten Einlagerungsbereichs, erstreckt sich über mehrere Jahrzehnte. Die Gesellschaft für Anlagen- und Reaktorsicherheit (GRS) hat für ein Tiefenlager in Salz eine Betriebsphase von 40 Jahren abgeschätzt (GRS 2012). Hieran schließt sich dann die Beobachtungsphase an, nach deren Ende der vollständige Verschluss des Tiefenlagers erfolgt. Bis zu diesem Zeitpunkt ist eine planmäßige Rückholung der eingelagerten Abfälle vorgesehen. Hierfür ist immer ein gesicherter Wissensstand über den Behälterzustand erforderlich. Somit ergibt sich für den Fall, dass direkt nach der Einlagerung des letzten Abfallgebindes mit dem Verschluss der Infrastrukturstrecken begonnen wird, ein Zeitrahmen von mehr als 40 Jahren, über den das Monitoring durchgeführt werden muss. Bei Planungen, die eine längere Beobachtungsphase für das Tiefenlager und eine längerfristige Option zur Rückholung vorsehen, kann sich dieser Zeitrahmen auf über 100 Jahre ausweiten. Aus der Forderung, dass die geotechnische Barriere nicht für Kabel durchörtert werden darf, folgt, dass alle direkt am Behälter positionierten Sensorsysteme über diesen Zeitraum zuverlässig arbeiten müssen.

Die Kernanforderungen an die Komponenten für das Behältermonitoring sind somit:

- Die Sensoren müssen in der Lage sein, die Messgröße auch in einer verfüllten Einlagerungsstrecke zuverlässig zu erfassen.
- Die Sensorsysteme müssen für die geplante Einsatzdauer von mehreren Jahrzehnten eine hohe Robustheit und Zuverlässigkeit besitzen.
- Die Sensorsysteme müssen entweder über eine eigene Energiequelle verfügen oder die erforderliche Energie für ihren Betrieb aus ihrem Umfeld generieren bzw. entnehmen können.
- Die Übertragung der Messwerte vom Sensorsystem an das Monitoringsystem muss drahtlos erfolgen.

Diese Anforderungen weichen stark von denen ab, die üblicherweise für Sensorsysteme im Anlagen- oder Bauwerksmonitoring gelten. In diesen Bereichen erfolgt ebenfalls ein betriebsbegleitendes Monitoringprogramm, das bei entsprechenden Nutzungszeiträumen die Entwicklung von Anlagen- bzw. Bauwerkskomponenten über Jahrzehnte hinweg erfasst. So konnte beispielsweise an der Talsperre Eibenstock die einwandfreie Funktion einbetonierter Ultraschallsensoren Jahrzehnte nach ihrem Einbau nachgewiesen werden (Niederleitinger et al. 2015). Da jeder Sensor eine gewisse Ausfallwahrscheinlichkeit hat und einer natürlichen Alterung unterliegt, werden Sensorsysteme im Anlagen- und Bauwerkmonitoring so gestaltet, dass eine regelmäßige Instandhaltung erfolgen kann. Durch die Möglichkeit, die Sensoren zu überprüfen, zu warten und gegebenenfalls auszutauschen, lassen sich in diesem Fall die Anforderungen an die langfristige Funktion der Sensorsysteme erheblich reduzieren. Die Forderung nach einem langfristig wartungsfreien Betrieb ist somit nur bei wenigen Anwendungen, beispielsweise in der Raumfahrttechnik, anzutreffen.

Für das Monitoring in einem Tiefenlager sind in den Bereichen Robustheit der Sensoren, Datenübertragung sowie Energieversorgung Lösungen erforderlich, die über den heutigen Stand der Technik hinausgehen. Aus diesem Grund ist das Monitoring im geotechnischen Umfeld in den letzten Jahren verstärkt in den Fokus der wissenschaftlichen Forschung gerückt. Forschungsprojekte wie „Monitoring Developments for Safe Repository Operation and Staged Closure – MoDeRn“ und das Nachfolgeprojekt „MoDeRn 2020“ haben sich in den vergangenen Jahren bereits mit dieser Thematik befasst und mögliche Lösungsansätze für Teilaspekte der oben genannten Aufgabenstellung aufgezeigt. Es konnte beispielsweise gezeigt werden, dass eine Signalübertragung per Funk durch die geologische Barriere prinzipiell möglich ist (Hart/Schröder 2012;

s. auch Jobmann/Liebenstund in diesem Band). Die Reichweite der Sender ist hierbei aber begrenzt. Diesem Umstand kann durch die Einrichtung einer mit Empfangstechnik ausgestatteten Monitoringstrecke oberhalb der Einlagerungsstrecke Rechnung getragen werden (Stahlmann et al. 2015). Auch für die Energieversorgung der Sensorsysteme wurden bereits einige Konzepte (z. B. Radioisotopenbatterien) vorgeschlagen. Diese orientieren sich an Techniken aus der Raumfahrt. Der Nachweis der Eignung für den Einsatz in einem Tiefenlager steht hierzu noch aus.

Ausgehend von den bisherigen Forschungsergebnissen erscheint es realistisch, dass in Zukunft sowohl für die Energieversorgung der Sensoren als auch für die Übertragung der Messdaten Systeme zur Verfügung stehen werden. Somit rückt die Forderung nach robusten Sensoren sowie Sensorsystemen in den Vordergrund. Da noch kein Endlagerbehälter existiert, ist es möglich, den Anforderungen an die Überwachung der technischen Barriere durch eine Integration von Sensorfunktionen in den Behälter zu begegnen.

Als Sensoren können Signalwandler verwendet werden, die auf unterschiedlichen physikalischen Prinzipien beruhen. Die Sensoren dürfen keine negativen Wechselwirkungen mit dem Umfeld induzieren, müssen für den Einsatz unter den gegebenen Randbedingungen geeignet sein (Temperatur, Druck, chemischer Angriff etc.) und mit geringen Energiemengen sicher funktionieren. Hinzu kommt die Forderung nach der Robustheit und Langzeitstabilität der Sensoren.

Zusätzlich sind redundante Messsysteme zur Erfassung von Messgrößen anzustreben. Durch den Einsatz von auf verschiedenen Messprinzipien beruhenden Sensoren können systematische Fehler eines einzelnen Messprinzips oder der spezifischen Sensoren identifiziert werden. Beispielsweise können für die Erfassung von Temperaturen zeitgleich Thermoelemente, Bimetallschalter und faseroptische Temperatursensoren verwendet werden. Hierdurch können selbstüberwachende Systeme realisiert werden, die es ermöglichen, die Messwerte einzelner Sensoren zu validieren und den Zustand des Sensorsystems zu ermitteln sowie die Interpretation der Daten erleichtern.

Wenn aufgrund des begrenzten Bauraums oder des Energiebedarfs für eine Messgröße keine redundanten Messsysteme vorgesehen werden können, spielt die Zuverlässigkeit des einzelnen Sensors die herausragende Rolle. Falls für eine Messgröße keine Sensoren zur Verfügung stehen, die die notwendige Zuverlässigkeit bei gleichzeitig geringem Energiebedarf aufweisen, ist eine kontinuierliche Erfassung dieser Messgröße nicht möglich. In diesem Fall kann es zielführend sein, Grenzwertschalter zu verwenden. Hierbei handelt es sich um robust aufgebaute Schalter, die auf einem störunempfindlichen physikalischen Prinzip, wie beispielsweise dem Phasenübergang eines Werkstoffs, beruhen. Mit Grenzwertschaltern ist nur eine diskontinuierliche Überwachung einer Messgröße möglich, da sie erst bei der Überschreitung eines Schwellenwerts auslösen. Sie ermöglichen allerdings durch eine materialinhärente Reversibilität eine sehr gute Zuverlässigkeit bei einer kompakten Baugröße. Einige Grenzwertschalter können zudem energieautark gestaltet werden. Bei Formgedächtnislegierungen zum Beispiel wandelt sich deren Gitterstruktur beim Überschreiten einer definierten Temperatur von einem Gittertyp in einen anderen um (Duering et al. 1990). Bei dieser Umwandlung wird Energie freigesetzt oder verbraucht, die mittels Technologien des sogenannten „Energy Harvesting" für eine kurzzeitige Funkübertragung des Sensorsignals über kurze Distanzen genutzt werden kann.

3 Zusammenfassung und Ausblick

Bei der geologischen Tiefenlagerung mit der Option der geplanten Rückholbarkeit der eingelagerten radioaktiven Abfälle stellt das Monitoring der technischen Barriere einen wichtigen Bestandteil der Endlagerplanung dar. Eine fundierte Kenntnis über den Zustand der Behälter ist für die Planung sowie die Durchführung einer Rückholung zwingend erforderlich. Eine Perforation der umgebenden geologischen und geotechnischen Barrieren muss vermieden werden. Die Sensoren und Sensorsysteme, die sich innerhalb dieser Barrieren befinden, sind weder für eine Wartung zugänglich, noch kann eine Energie- und Datenübertragung zu den Sensoren mittels einer Verkabelung von außen realisiert werden. Des Weiteren existieren für viele der relevanten Parameter des Behälters noch keine Sensoren, die eine Erfassung von Messdaten innerhalb eines geologischen Tiefenlagers ermöglichen. Die größten technischen Herausforderungen beim Monitoring der technischen Barriere bestehen bei der Energieversorgung der Sensoren, der Datenübertragung sowie der Entwicklung geeigneter Sensoren und Sensorsysteme. Für alle diese Fragestellungen sind derzeit noch keine Lösungen verfügbar, deren Eignung für den Einsatz im geologischen Tiefenlager ausreichend nachgewiesen worden sind. Die Ergebnisse aktueller Forschungsvorhaben wie beispielsweise MoDeRn zeigen, dass voraussichtlich sowohl für die Energieversorgung als auch für die Datenübertragung über kurze Strecken in absehbarer Zukunft Lösungen entstehen werden, mit denen energieeffiziente Sensorsysteme betrieben werden können (Jobmann et al. 2012).

Weiterer Forschungsbedarf besteht bei der Entwicklung langzeitstabiler Sensorsysteme, die eine hohe Zuverlässigkeit sowie eine geringe Ausfallwahrscheinlichkeit besitzen. Hierfür sind zum einen möglichst stabile, langlebige Einzelsensoren und zum anderen fehlertolerante Sensorsysteme zu entwickeln. Beispielsweise können durch die Auswertung mehrerer redundanter Sensoren selbstüberwachende Systeme realisiert werden. Diese können Alterungseffekte sowie Defekte an den Sensoren identifizieren und somit die Verlässlichkeit der ermittelten Daten erheblich steigern.

Aufgrund der Komplexität und Vielzahl der unterschiedlichen für das Behältermonitoring relevanten Parameter ist davon auszugehen, dass nicht für jede Messgröße ausreichend zuverlässige Sensoren entwickelt werden können, die sich für den Einsatz unter den gegebenen Bedingungen eignen. In diesem Fall ist zu untersuchen, ob gegebenenfalls Grenzwertschalter entwickelt werden können, mit denen sich diese Messgröße diskret erfassen lässt.

Zur Implementierung eines Monitoringkonzepts speziell für die technische Barriere in einem geologischen Tiefenlager sind somit noch viele Lösungen für einzelne Teilaspekte zu entwickeln. Dennoch werden sich voraussichtlich nicht alle relevanten Messgrößen der Behälter zuverlässig *in situ* bestimmen lassen. Im Zuge der Planung der Rückholoption macht es Sinn, den Behälter als technische Barriere in das Monitoringkonzept einzubeziehen. Durch die im Forschungsverbund ENTRIA entwickelten ENCON-Behältersysteme für die unterschiedlichen Wirtsgesteine sind Grundlagen geschaffen worden, welche hinsichtlich der Integration von Monitoringfunktionen in den Behälter in naher Zukunft weiterentwickelt werden müssen (Hassel et al. 2019). Mit diesen z. B. „smart-EnCon“-Behälterkonzepten besteht die Möglichkeit, die technische Barriere besser zu überwachen und der bisherigen Black-Box, die der Behälter in allen bisherigen Konzepten darstellt, eine intelligente Funktion zu verleihen. Damit kann die geforderte Sicherheit der Endlagerung wesentlich verbessert und das Risiko von Fehlentscheidungen minimiert werden.

Literatur

BMU – Bundesministerium für Umwelt, Naturschutz, Bau und Reaktorsicherheit (2010): Sicherheitsanforderungen an die Endlagerung wärmeentwickelnder radioaktiver Abfälle. Berlin

Bohner, E. (2014): Reflections on the use of monitoring technologies in geological nuclear waste repositories. GeoRepNet meeting, Nottingham

Brasser, T.; Miehe, R. (2008): Endlagerung wärmeentwickelnder radioaktiver Abfälle in Deutschland. Anhang Verfüllen: Technische Verfüll- und Verschlussmaßnahmen in einem Endlager; GRS-247/21, Braunschweig/Darmstadt

Duering, T.W.; Melton, K.N.; Stöckel, D.; Wayman, C.M. (1990): Engineering Aspects of Shape Memory Alloys. London

GRS – Gesellschaft für Anlagen- und Reaktorsicherheit (2012): Endlagerauslegung und -optimierung. Bericht zum Arbeitspaket 6: Vorläufige Sicherheitsanalyse für den Standort Gorleben. GRS-281, Braunschweig

Hart, J.; Schröder, T.J. (2012): Wireless Transmission of Monitoring Data out of the HADES Underground Laboratory. WM2012 Conference, February 26 – March 1, 2012, Phoenix, AZ

Hassel, T.; Köhler, A.; Kirt, Ö.S. (2019): Das ENCON-Behhälterkonzept. Generische Behältermodelle zur Einlagerung radioaktiver Reststoffe für den interdisziplinären Optionenvergleich. ENTRIA-Arbeitsbericht 16, Hannover

Hofmeister, J. (2015): Vom Lagerbecken zum Standortzwischenlager. Biblis

Jobmann, M.; Schröder, T.J.; White, M. (2012): Development of a Theoretical Monitoring System Design for a HLW Repository Based on the „MoDeRn Monitoring Workflow" (A Case Study) – 12044. WM2012 Conference, February 26 – March 1, 2012, Phoenix, AZ

Müller-Lyda, I. (1999): Eigenschaften von Salzgrus als Versatzmaterial im Wirtsgestein Salz. Workshop Entsorgung, GRS-143, Köln

Niederleitinger, E.; Krompholz, R.; Müller, S. (2015): 36 Jahre Talsperre Eibenstock – 36 Jahre Überwachung des Betonzustandes durch Ultraschall. 38. Dresdener Wasserbaukolloquium – Messen und Überwachen im Wasserbau und am Gewässer, Dresden

Solente, N.; Bergmans, A.; Sineriz, J-L.; Breen, B.; Jobmann, M. (2013): Overview of the MoDeRn project: A reference framework for developing a monitoring program. In: Monitoring in geological disposal of radioactive waste – Objectives, Strategies, Technologies and Public Involvement. Proceedings, Appendix C. Full Conference Papers, Luxemburg, S. 10–26

Stahlmann, J.; Mintzlaff, V.; Leon Vargas, R. (2015): Generische Tiefenlagermodelle mit Option zur Rückholung der radioaktiven Reststoffe: Geologische und Geotechnische Aspekte für die Auslegung. ENTRIA-Arbeitsbericht-03, Braunschweig

StandAG – Standortauswahlgesetz (2017): Gesetz zur Suche und Auswahl eines Standortes für ein Endlager für hochradioaktive Abfälle. Bearbeitungsstand Mai 2017

Voßnacke, A.; Klein, K.; Kühne, B. (2004): CASTOR® HAW28M – a high heat load cask for transport and storage of vitrified high level waste containers, 14th International Symposium on the Packaging and Transportation of Radioactive Materials (PATRAM), Berlin

Armin Grunwald

Die Endlagerkommission des Deutschen Bundestages: Prozess- und Endlagermonitoring

1 Die Endlagerkommission

Die nach der Havarie von Fukushima im Jahre 2011 beschlossene Beendigung der Kernenergienutzung in Deutschland ermöglichte den Neustart der Endlagersuche in einem breiten Konsens von CDU/CSU bis zu den Grünen. Lediglich DIE LINKE und Teile der Anti-Atom-Bewegung verweigerten sich. Auf dieser breiten Basis hat der Bundestag 2013 mit dem Standortauswahlgesetz die Grundlagen für die Arbeit der Endlagerkommission gelegt (korrekter Name: Kommission Lagerung hoch radioaktiver Abfallstoffe). Danach sollen im Sinne des Verursacherprinzips die im Inland verursachten radioaktiven Abfälle in Deutschland sicher gelagert werden.

Aufgabe der Kommission war die Entwicklung eines Standortauswahlverfahrens, die Einigung auf Entscheidungskriterien im Auswahlprozess und die Konzeption einer Institutionenstruktur mit Beteiligung der Öffentlichkeit. Ihr gehörten 34 Mitglieder an: neben den beiden Vorsitzenden jeweils acht Vertreter aus Bundestag, Bundesrat, Wissenschaft und Gesellschaft (je zwei Vertreter aus Wirtschaft, Gewerkschaft, Kirchen und Umweltverbänden). Mit dieser Zusammensetzung sollten alle wichtigen gesellschaftlichen Positionen einbezogen werden, um einerseits eine breite Zustimmung bei dieser schwierigen und konfliktreichen Aufgabe zu erreichen und andererseits durch die starke Beteiligung von Bundestag und Bundesrat gute Bedingungen für die Übernahme ihrer Ergebnisse in die politische Umsetzung zu schaffen. Die Kommission gab wichtige Impulse für einen in dieser Form neuen Prozess, der mit Smeddinck als sanfte Regulierung verstanden werden kann (Smeddinck 2019).

Die Kommission hat in maximaler Transparenz gearbeitet. Videomitschnitte und Wortprotokolle liegen vor und können eingesehen werden (EK 2016). Nach über zweijähriger Arbeit mit 35 Plenarsitzungen und knapp 100 Arbeitsgruppensitzungen, Anhörungen und Workshops sowie zahlreichen Gutachten und Arbeitspapieren hat die Kommission am 5. Juli 2016 ihren Abschlussbericht „Verantwortung für die Zukunft – ein faires und transparentes Verfahren für die Auswahl eines nationalen Endlagerstandorts“ vorgelegt (EK 2016). Der Bundestag hat auf Grundlage der Vorschläge der Endlagerkommission das Standortauswahlgesetz (StandAG) 2017 in Kraft gesetzt. Rasch wurden die vorgesehenen Institutionen geschaffen: das Bundesamt für die Sicherheit der nuklearen Entsorgung (BASE; zunächst Bundesamt für kerntechnische Entsorgungssicherheit BfE), die Bundesgesellschaft für Endlagerung (BGE) und das Nationale Begleitgremium (NBG) mit der Geschäftsstelle am Umweltbundesamt (UBA). Im Herbst 2017 wurde das Standortauswahlverfahren gestartet (StandAG 2017) und wird seitdem gemäß den von der Endlagerkommission entworfenen Prozessschritten abgearbeitet. Unerwartete Ereignisse wie die überraschend große Zahl der im ersten Schritt für die weitere Untersuchung identifizierten Teilgebiete (s. u.) und die Corona-Pandemie, welche die Beteiligungsmöglichkeiten auf digitale Formate beschränkte, waren bereits erste Stressfaktoren für das Verfahren.

2 Auf dem Weg zu einem Standort mit der bestmöglichen Sicherheit[1]

Die Kommission empfiehlt nach einer umfassenden Beschäftigung mit einer Vielzahl von Optionen die Verbringung der Abfälle in ein Endlagerbergwerk in einer tiefen geologischen Formation. Die Langzeitsicherheit soll zu einem großen Teil solchen geologischen Formationen überantwortet werden, die – anders als gesellschaftliche Strukturen – die erforderliche Langzeitstabilität aufweisen. In Deutschland kommen hierfür Salz, Ton und Kristallin als Wirtsgesteine in etwa 300–800 Metern Tiefe infrage. Konzeptionell neu ist die Forderung nach Reversibilität einmal getroffener Entscheidungen im Sinne eines lernenden Verfahrens, vor allem um Fehlerkorrekturen zu ermöglichen. Die Kommission definiert den gesuchten Standort wie folgt:

> Der gesuchte Standort für ein Endlager insbesondere für hoch radioaktive Abfallstoffe bietet für einen Zeitraum von einer Million Jahren die nach heutigem Wissensstand bestmögliche Sicherheit für den dauerhaften Schutz von Mensch und Umwelt vor ionisierender Strahlung und sonstigen schädlichen Wirkungen dieser Abfälle. Dieser Standort ist nach den entsprechenden Anforderungen in einem gestuften Verfahren durch einen Vergleich zwischen den in der jeweiligen Phase geeigneten Standorten auszuwählen. Lasten und Verpflichtungen für zukünftige Generationen sind möglichst gering zu halten. Geleitet von der Leitidee der Nachhaltigkeit wird der Standort mit der bestmöglichen Sicherheit nach dem Stand von Wissenschaft und Technik mit dem in diesem Bericht beschriebenen Auswahlverfahren und den darin angegebenen und anzuwendenden Kriterien und Sicherheitsuntersuchungen festgelegt. Während des Auswahlverfahrens und später am gefundenen Standort muss eine Korrektur von Fehlern möglich sein (EK 2016, S. 24).

---

1 Diese Darstellung orientiert sich an Grunwald (2016) und übernimmt einige Passagen textidentisch.

Dieser Standort mit der bestmöglichen Sicherheit muss sich auf transparente und nachvollziehbare Weise als Ergebnis des Standortauswahlverfahrens ergeben. Damit ist das Problem der Standortsuche prozeduralisiert: Der bestgeeignete Standort ergibt sich durch das Durchlaufen des Verfahrens. Der Standort wird weder szientistisch ausgerechnet noch politisch festgelegt, sondern zum Abschluss einer Reihe von wissenschaftlichen Untersuchungen und transparenten Abwägungsschritten bestimmt, von denen jeder einzelne transparent und nachvollziehbar sein soll. Abkürzungsmöglichkeiten im Verfahren sind nicht vorgesehen bzw. würden dem Ansatz widersprechen. Wiederkehrende Bekundungen zur Eignung oder Nicht-Eignung bestimmter Orte, Regionen oder ganzer Bundesländer sind ohne sachlichen Belang, sondern stellen politische Bekundungen dar. Erst im Verfahren selbst kann sich zeigen, welche Standorte wie geeignet sind. Für den Neubeginn des Verfahrens nach über vierzig Jahren einer von Protest, Intransparenz, hoheitlichen Durchsetzungsversuchen des Standorts Gorleben und einem erheblichen Vertrauensverlust in staatliche Institutionen und Prozesse gekennzeichneten Entwicklung wurde symbolisch Deutschland als „weiße Landkarte" bezeichnet. Auf dieser soll zunächst jeder Ort einschließlich Gorleben als potenzieller Standort für ein Endlager gleich betrachtet werden. Erst im Verlauf des Auswahlverfahrens soll wissenschaftsbasiert nach von der Endlagerkommission empfohlenen und im Anhang des StandAG niedergelegten Kriterien die weiße Landkarte Schritt für Schritt eingegrenzt werden, bis zum Schluss nach vergleichenden Analysen der Standort mit der bestmöglichen Sicherheit identifiziert wird.

Das Endlagerbergwerk in einer tiefen geologischen Formation soll in einer mehr oder weniger fernen Zukunft verschlossen werden, um keine Belastungen der belebten Umwelt und zukünftiger Generationen zu verursachen. Alle Schritte bis zu diesem Zustand müssen zu Beginn des Verfahrens plausibel dargestellt werden, um die Erwartung zu begründen, auf diesem Weg eine nachhaltige, verantwortliche und sichere Lösung zu ermöglichen. Hierfür ist zunächst in einem vergleichenden Auswahlverfahren der Standort mit der bestmöglichen Sicherheit zu bestimmen. Danach kann das Endlagerbergwerk geplant, genehmigt und gebaut werden. Es schließt sich die Phase der Einlagerung der Abfälle an, gefolgt von einem Zeitraum der Beobachtung der geologischen Entwicklung wesentlicher Parameter im Bergwerk (wie z. B. Temperatur und Gasdruck). So lange sollen die Abfallbehälter jederzeit rückholbar sein. Sobald hinreichende Evidenz besteht, dass sich das Endlagerbergwerk in eine sichere Richtung entwickelt, kann es verschlossen werden. Die Behälter sollen so ausgelegt werden, dass die Abfälle aber auch noch Jahrhunderte später geborgen werden könnten, falls hierzu ein Anlass besteht.

Dieser Ansatz bürdet dem Standortauswahlverfahren und den dabei zum Einsatz kommenden Kriterien die zentrale Last auf, damit das Ergebnis der

Suche auch den Erwartungen entspricht. Damit bilden geowissenschaftliche Ausschlusskriterien, Mindestanforderungen und Abwägungskriterien die substanzielle Mitte des Auswahlverfahrens. Ihnen kommt zusammen mit standortbezogenen Sicherheitsuntersuchungen die höchste Bedeutung zu, den Suchprozess schrittweise, vergleichend und nachprüfbar zum Standort mit der bestmöglichen Sicherheit zu navigieren. Dieser Kriteriensatz wird über die Laufzeit des Auswahlverfahrens konstant gehalten, um Verzerrungen zu vermeiden. Ausgehend von der „weißen Landkarte" Deutschlands werden zunächst Regionen ausgeschlossen, die geologisch nicht in Frage kommen (z. B. Gegenden mit Vulkanismus). Unter den verbleibenden Regionen wird dann schrittweise mittels der Abwägungskriterien vergleichend ermittelt, welche so vielversprechend erscheinen, dass sie zunächst obertägig, dann eine verkleinerte Auswahl auch untertägig erkundet werden sollen. Unter diesen wird letztlich der Standort mit der bestmöglichen Sicherheit bestimmt.

Dieses Verfahren bedarf eines Höchstmaßes an Transparenz und Qualitätssicherung, muss sich wissenschaftlichem Review, internationalem Vergleich und öffentlicher Diskussion stellen sowie die vorgeschlagenen weitreichenden Beteiligungsmöglichkeiten umsetzen. Im Standortauswahlgesetz wurde der Anspruch an das Verfahren in der Zweckbestimmung wie folgt (StandAG § 1[2]) festgehalten:

> Mit dem Standortauswahlverfahren soll in einem partizipativen, wissenschaftsbasierten, transparenten, selbsthinterfragenden und lernenden Verfahren für die im Inland verursachten hochradioaktiven Abfälle ein Standort mit der bestmöglichen Sicherheit für eine Anlage zur Endlagerung nach § 9a Absatz 3 Satz 1 des Atomgesetzes in der Bundesrepublik Deutschland ermittelt werden.

Das vom Bundestag in Abstimmung mit dem Bundesrat berufene Nationale Begleitgremium wurde bereits vor Beginn des Standortauswahlverfahrens zum Jahresende 2016 installiert, um das gesamte Verfahren im Blick zu behalten und die Einhaltung der Ideale des Verfahrens in seiner praktischen Umsetzung zu überwachen (NBG 2021). Regionalkonferenzen sollen in naher Zukunft Nachprüfrechte bei allen Entscheidungsschritten zugesprochen bekommen; sie sollen dabei mit Ressourcen ausgestattet werden, diese Rechte umzusetzen. Im Rahmen eines „selbsthinterfragenden Systems" soll auf die Früherkennung von Fehlern und die Fehlerkorrektur geachtet werden, sollen aber auch mögliche andere Lösungen der Endlagerherausforderung beobachtet werden, um ggf. das ursprünglich vorgesehene Verfahren modifizieren oder gar umsteuern zu können (vgl. Teil 3).

Hiermit wird ein langer Weg vorgezeichnet, der bis zum Verschluss des Bergwerks mehr als nur einige Jahrzehnte benötigen wird. Datenerhebungen, Erkundungen möglicher Standorte, Genehmigungsverfahren, mögliche rechtliche Auseinandersetzungen, aber auch die Notwendigkeit der Sorgfalt in den Abwägungen, die Umsetzung der Nachprüfrechte der Öffentlichkeit und weitere Beteiligungsformate benötigen erhebliche Zeit. Weil jedoch sehr lange Zeiträume nachfolgende Generationen erheblich belasten und umfangreiche Zwischenlagerungen notwendig machen würden, auch weil damit die Gefahr des Erlahmens und Ermüdens steigen würde, muss darauf hingearbeitet werden, dass der Gesamtprozess in einem vertretbaren Zeitrahmen verbleibt. Das Standortauswahlgesetz sieht die Festlegung des Endlagerstandorts für das Jahr 2031 vor, während die Endlagerkommission erhebliche Zweifel an der Realisierbarkeit eines derart schnellen Verfahrens geäußert hatte. In dem Zielkonflikt zwischen bestmöglicher Sicherheit und substanzieller Öffentlichkeitsbeteiligung auf der einen Seite und dem Wunsch nach Schnelligkeit im Verfahren auf der anderen Seite haben Sicherheit und Beteiligung Priorität (EK 2016, 248ff.), wie dies zurzeit auch das Nationale Begleitgremium vertritt.

3 Das lernende Verfahren und der Bedarf an Monitoring

Von Beginn der Kommissionsarbeit an spielte der Gedanke der Reversibilität von Entscheidungen im Rahmen eines lernenden Verfahrens eine zentrale Rolle. Im Gegensatz zu früheren Vorstellungen, die Abfälle möglichst rasch und *de facto* unumkehrbar in ein Endlagerbergwerk zu verbringen, hat die Endlagerkommission die Möglichkeiten des Lernens, der Fehlerkorrektur und der Umsteuerung sogar in ihr Leitbild aufgenommen (EK 2016, S. 23):

> Die Kommission versteht ihre Arbeit und die spätere Standortsuche als ein *lernendes Verfahren.* Dabei sind Entscheidungen gründlich auf mögliche Fehler oder Fehlentwicklungen zu prüfen. Möglichkeiten für eine spätere Korrektur von Fehlern sind vorzusehen. Auch deshalb ist die Öffentlichkeit an der Suche von Anfang an breit zu beteiligen. Ziel ist ein offener und pluralistischer Diskurs. Vor der eigentlichen Standortsuche müssen Entsorgungspfad und Alternativen, grundlegende Sicherheitsanforderungen, Auswahlkriterien und Möglichkeiten der Fehlerkorrektur wissenschaftsbasiert und transparent entwickelt, genau beschrieben und öffentlich debattiert sein. Bei einem späteren Umsteuern oder einer späteren Korrektur von Fehlern muss dies ebenfalls gewährleistet sein.

An einer weiteren Stelle heißt es (EK 2016, S. 31.):

> Konzeptionell neu ist die an zukunftsethischen Prinzipien und dem Wunsch nach weitgehenden Möglichkeiten der Fehlerkorrektur ausgerichtete Forderung nach Reversibilität einmal getroffener Entscheidungen im Sinne eines lernenden Verfahrens, um das Ziel der bestmöglichen Sicherheit [...] zu erreichen. Reversibilität, also die Möglichkeit zur Umsteuerung im laufenden Verfahren, ist erforderlich, um Fehlerkorrekturen zu ermöglichen, um Handlungsoptionen für zukünftige

Generationen offenzuhalten, zum Beispiel zur Berücksichtigung neuer Erkenntnisse, und kann zum Aufbau von Vertrauen in den Prozess beitragen. Konzepte der Rückholbarkeit oder Bergbarkeit der Abfälle beziehungsweise der Reversibilität von Entscheidungen sind dafür zentral.

Diese geradezu überdeutliche Orientierung an Prinzipien einer Gestaltung des Weges hin zu einer Endlagerung ist sicher erklärungsbedürftig. Hier dürften eine Rolle gespielt haben:

- Erfahrungen mit der Asse, wo auf Rückholbarkeit und Fehlerkorrektur kaum Wert gelegt wurde, mit entsprechenden Konsequenzen. Es gab kaum eine Kommissionssitzung, in der die Asse nicht erwähnt wurde (vgl. zum Hintergrund Hocke et al. 2016). Der folgende Passus aus dem Bericht macht diese Motivation deutlich (EK 2016, S. 190):

  > Aus dem Scheitern der Endlagerung radioaktiver Abfälle im ehemaligen Salzbergwerk Asse ergeben sich nach Auffassung der Kommission Konsequenzen für den Umgang mit abweichenden wissenschaftlichen Meinungen. Frühe Warnungen vor Zuflüssen in das Bergwerk Asse blieben seinerzeit ohne Konsequenzen und hatten auch negative Folgen für warnende Wissenschaftler. Bei der Schachtanlage Asse hätte man einen falschen Weg früher korrigieren können, wenn man kritische Stimmen frühzeitig ernst genommen hätte. Je später man einen Fehler erkennt, desto teurer kann eine Korrektur werden. Die Geschichte der Schachtanlage zeigte zudem, wie unerlässlich eine vom Betreiber unabhängige Begutachtung ist.

- Geringes Vertrauen in die langfristige Planbarkeit komplexer Prozesse wie der Suche nach einem Endlager und seiner konzeptionellen wie technischen Ausgestaltung.
- Erwartungen, dass während der langen Dauer des Verfahrens neue wissenschaftliche Erkenntnisse gewonnen werden könnten, die zu einer Modifikation Anlass geben könnten.
- Ethische Argumente dahingehend, dass zukünftige Generationen zwar einerseits möglichst nicht mit unseren Hinterlassenschaften belastet werden sollten, dass ihnen aber die Möglichkeit zum Treffen eigenständiger Entscheidungen auf Basis eigener Wert- und Prioritätensetzungen offenzuhalten sei.

Ein lernendes Verfahren ist grundsätzlich auf Monitoring angewiesen (Hocke et al. 2012): Es muss über sich selbst Bescheid wissen, um mögliche Probleme zu erkennen, genauso wie es seine Umwelt sorgfältig beobachten muss, um Entwicklungen zu entdecken, die möglicherweise Einfluss auf den weiteren Gang des Verfahrens haben könnten. Lernen muss in dieser Spannung zwischen innen und außen erfolgen. Beides erfordert geeignete Beobachtungs- und Reflexionsinstrumente.

4 Prozess- und Endlagermonitoring[2]

Das Standortauswahlverfahren und die sich daran anschließenden Phasen der Errichtung des Endlagerbergwerks, des Baus der erforderlichen Logistik und der Einlagerung der Abfälle bis zu einem Verschluss des Bergwerks werden viele Jahrzehnte in Anspruch nehmen und erfordern eine *Long-Term Governance* (Kuppler/Hocke 2019). Auch nach Verschluss soll der Standort noch über Jahrhunderte unter Beobachtung verbleiben. Diese selbst für andere technische Großprojekte extrem lange Zeitdauer des Gesamtvorgangs macht einen klassischen Planungsansatz unmöglich, sondern erfordert, den Prozess selbst auch von Anfang an lernend zu gestalten und für diesen Zweck einem begleitenden Monitoring und einer periodischen und kritischen Evaluierung zu unterziehen:

> Die Kommission empfiehlt aus heutiger Sicht, den gesamten Endlagerprozesses [sic!] als sich selbsthinterfragendes System zu gestalten und über kontinuierliches Prozessmonitoring Fehler und unerwünschte Entwicklungen möglichst zu vermeiden (EK 2016, S. 190).

Unter *Prozessmonitoring* versteht die Kommission das begleitende Monitoring des gesamten Prozessweges hin zum Verschluss des Endlagers, aller dabei stattfindenden Entscheidungsprozesse und ihrer Folgen sowie auftretender Herausforderungen, aber auch der relevanten Veränderungen im Umfeld. Zu Letzteren gehören politische Veränderungen, Wertewandel, neue wissenschaftliche Erkenntnisse oder technologische Optionen. Im Verfahren selbst sollen die Ergebnisse dieser Beobachtungen im Hinblick auf die nächsten Schritte analysiert und reflektiert werden, um möglichen Umsteuerungsbedarf, Umsteuerungsmöglichkeiten oder Adjustierungschancen zu erkennen und im weiteren Verlauf berücksichtigen zu können. Ein derartiges Prozessmonitoring (EK 2016, S. 273 ff.) muss bereits mit Beginn des Auswahlverfahrens einsetzen und zumindest folgende Aspekte[3] umfassen:

- regelmäßige Reflexion und Bewertung des Standes des Verfahrens gemessen an den selbst gesetzten Zielen; möglicherweise Modifikation der Ziele und der vorgesehenen Zeitspannen;

2 Die Darstellung folgt inhaltlich dem Kap. 6.3.6 des Berichts der Endlagerkommission (EK 2016).

3 Siehe hierzu insbes. EK 2016, S. 274.

- regelmäßige Evaluierung der institutionellen Situation: Betreiber, Behördenstruktur, Aufsicht, Transparenz etc.;
- regelmäßige Erhebung der Haltung in der Bevölkerung zum Prozess der Endlagerung zur möglichst frühzeitigen Aufdeckung von Vertrauensproblemen und von Schwachstellen der Beteiligung;
- Reflexion der Frage, welche Parameter für ein Monitoring beobachtbar sind oder beobachtet werden sollen und welche Verfahren hierfür eingesetzt oder erst entwickelt werden sollen;
- regelmäßige Prüfung ob die für die Erkundung eingesetzte Vorgehensweise sowie die vorgesehene Technik dem nationalen und internationalen Stand von Wissenschaft und Technik entsprechen;
- regelmäßige Prüfung des Wissensstandes bei anderen potenziellen Entsorgungspfaden.

Der lange Zeitraum und die Offenheit zukünftiger Entwicklungen bringen mit sich, dass zukünftig einzusetzende Monitoring- und Erkundungsmethoden zum jetzigen Zeitpunkt noch nicht festgelegt werden können, sondern selbst im Laufe des Verfahrens erprobt, reflektiert und angepasst werden müssen, unter Berücksichtigung des jeweils geltenden internationalen Standes von Wissenschaft und Technik.

Sobald das Endlager beladen ist, muss die Entwicklung des geologisch-technischen Gesamtsystems mit den enthaltenen Abfällen und deren Wirkungen beobachtet werden. Diese Verpflichtung hat die Endlagerkommission als *Endlagermonitoring* bezeichnet. Es dient dem Zweck, „den Zustand der geologischen Formation, der hydrogeologischen Verhältnisse und der Abfälle […] systematisch zu beobachten" (EK 2016, S. 275 f.). Auf diese Weise sollen mögliche Fehlentwicklungen oder unvorhergesehene Verläufe frühzeitig entdeckt werden, um ggf. daraus Konsequenzen ziehen zu können, im Extremfall bis hin zur Rückholung oder Bergung der Abfälle. Monitoring-Einrichtungen sollten beispielsweise Spannungszustände und ihre Entwicklung oder die Bildung potenzieller Wasserdurchlässigkeit überwachen, die Temperaturentwicklung beobachten, einen Wasserzutritt sofort detektieren und über Gasbildung oder eine Radionuklidfreisetzung in den Nahbereich hinein informieren.

Das Monitoring der geologischen Formation für sich sollte bereits mit der Festlegung des Standortes beginnen, um mögliche durch Bergwerk und Abfälle verursachte Veränderungen bestmöglich erkennen zu können (EK 2016, S. 275). Wenn auch das Monitoring der soziotechnischen Konstellation aus Geologie und Endlagerbergwerk technisch erst mit der Standortbestimmung konkret geplant werden kann, muss vorher festgelegt werden, welche Parameter zu beobachten sind, da diese Auswirkungen auf die Auslegung der erforderlichen Technologien (Sensoren und Datenübertragung an die Oberfläche) und auf die Ingangsetzung entsprechender Forschung und Entwicklung haben, damit die Technologien auch verfügbar sind, wenn sie zum Einsatz kommen sollen. Das Endlagermonitoring ist mindestens bis zum Abschluss der Phase der Bergbarkeit der Abfälle (gegenwärtig geplant: etwa 500 Jahre nach Verschluss des Bergwerks) erforderlich.

Das Endlagermonitoring steht jedoch in einem Zielkonflikt. Einerseits geht es darum, die sicherheitsrelevanten Parameter für ein Endlager möglichst vollständig und engmaschig zu überwachen. Andererseits jedoch werden mit eingebauten Sensoren/Messgeräten und vor allem mit den damit verbundenen Datenleitungen auch potenzielle Schwachstellen für Undichtigkeiten und Wasserzutritt geschaffen. Dieser Konflikt tritt verschärft nach Verschluss des gesamten Bergwerks auf, da das Monitoring dann noch für Jahrhunderte weitergeführt werden soll. Technische Entwicklungen zur kabellosen Datenübertragung mit entsprechend neuen Monitoring-Möglichkeiten würden hier für Abhilfe sorgen.

Das Endlagermonitoring wird während des gesamten Prozesses parallel zu den Etappen der Endlagerung entwickelt und weiterentwickelt. Zu unterschiedlichen Zeitpunkten sind unterschiedliche Informationen relevant, die hinsichtlich ihrer Bedeutung für die kurzfristige und dauerhafte Sicherheit des Endlagers bewertet werden müssen. Damit kann die Verlässlichkeit des Endlagersystems ständig geprüft und die technisch-wissenschaftliche Entscheidungsgrundlage zur Problem- und Fehlererkennung bereitgestellt werden. Hierfür wiederum müssen transparente Maßstäbe entwickelt werden, um das Ausmaß und die Ausprägung von Abweichungen vom erwarteten Verlauf daraufhin interpretieren zu können, ob das Ergreifen von Fehlerkorrektur- oder Zusatzmaßnahmen erforderlich wird.

5 Standortauswahlgesetz und Nationales Begleitgremium

Die Endlagerkommission hat also dem Gedanken eines lernenden Verfahrens hohe Priorität eingeräumt, um das Ziel einer sicheren und wartungsfreien Endlagerung schrittweise mit den Wünschen nach Reversibilität von Entscheidungen, Rückholbarkeit der Abfälle, Ermöglichung von Fehlerkorrekturen und Lernmöglichkeiten im Prozess zu verbinden. Zumindest bis zur Erreichung des Endzustandes des nach diesen Anforderungen gestalteten Entsorgungspfades müssen Vorkehrungen für eine permanente Überprüfung des Entsorgungsprozesses unter dem Blickwinkel von Sicherheit, Transparenz und Beteiligung getroffen werden. Um die Notwendigkeit von Umsteuerungen im Prozess, z. B. zur Fehlerkorrektur, überhaupt erkennen zu können, bedarf es entsprechend geeigneter Formen des Monitorings. Das gilt insbesondere für einschneidende Schritte im Entsorgungsprozess, aber auch für einschneidende gesellschaftliche

Veränderungen. Weiterhin bedarf es institutioneller Vorkehrungen und Zuständigkeiten, damit das Erbe der Endlagerkommission nicht in rhetorischen Worthülsen endet.

Im StandAG wird das Endlagermonitoring nicht explizit angesprochen, da es sich nur auf den Prozess der Standortsuche erstreckt. Das Prozessmonitoring dagegen taucht zwar nicht als Begriff, aber der Sache nach unverkennbar auf, wenn im Gesetz von einem lernenden Verfahren und selbsthinterfragenden System gesprochen wird. Die Umsetzung wird von den beteiligten Institutionen erwartet mit besonderer Betonung des Nationalen Begleitgremiums (NBG 2021):

> Aufgabe des pluralistisch zusammengesetzten Nationalen Begleitgremiums ist die vermittelnde und unabhängige Begleitung des Standortauswahlverfahrens, insbesondere der Öffentlichkeitsbeteiligung, mit dem Ziel, so Vertrauen in die Verfahrensdurchführung zu ermöglichen. Es kann sich unabhängig und wissenschaftlich mit sämtlichen Fragestellungen das Standortauswahlverfahren betreffend befassen, die zuständigen Institutionen jederzeit befragen und Stellungnahmen abgeben. Es kann dem Deutschen Bundestag weitere Empfehlungen zum Standortauswahlverfahren geben (StandAG 2017 § 8[1]).

Das NBG ist pluralistisch zusammengesetzt. Dem Gremium gehören Persönlichkeiten des öffentlichen Lebens an, die von Bundestag und Bundesrat gewählt werden, und Bürgervertreter*innen, die in einem vom Bundesumweltministerium initiierten Beteiligungsverfahren ermittelt werden. Das NBG ist unabhängig, also nicht weisungsgebunden. Es kann sich mit sämtlichen Themen in Bezug auf das Standortauswahlverfahren befassen, die zuständigen Institutionen befragen und hat das Recht auf Akteneinsicht.

Das Prozessmonitoring als Grundlage für ein lernendes und selbsthinterfragendes Verfahren war im NBG in gewisser Weise seit Beginn seiner Tätigkeit durch die kritische Begleitung der Verfahrensschritte, etwa in Bezug auf Öffentlichkeitsbeteiligung, präsent. Hier ist beispielsweise die Begleitung der Fachkonferenz Teilgebiete zu nennen, die durch die Corona-Pandemie vor die Situation gestellt wurde, nur Online-Formate für die Beteiligung verwenden zu dürfen. Das NBG wird die Vor- und Nachteile von Online-Beteiligung im Standortauswahlverfahren durch ein Gutachten evaluieren lassen und sich auf dieser Basis eine Meinung zur möglichen Nutzung von Online-Formaten auch in der Zukunft zu bilden. Ebenfalls hat sich das NBG vorgenommen, in den Zeiten zwischen den drei Terminen der Fachkonferenz als Ort des Lernens und der Reflexion zu fungieren.

Im Jahr 2021 hat das NBG beschlossen, zur vertieften Behandlung dieser Thematik eine eigene Fachgruppe zu gründen (FG IV 2021). Sie ist zurzeit dabei, institutionelle Orte gegenseitigen und gemeinsamen Lernens zu erkunden.

Auf diese Weise hat das NBG das Prozessmonitoring explizit zu seiner Kernaufgabe gemacht. Parallel dazu hat das NBG entschieden, das Standortauswahlverfahren durch ein unabhängiges und internationales Peer Review untersuchen zu lassen, das selbstverständlich auch die eigene Rolle umfassen soll. Durch die Begutachtung des bisherigen Verfahrens und des aktuellen Herangehens im Hinblick auf die fünf Prinzipien des StandAG – partizipativ, wissenschaftsbasiert, transparent, selbsthinterfragend und lernend – durch unabhängige internationale Expertinnen und Experten soll größtmögliche Transparenz erreicht und Vertrauen in das Verfahren gesteigert werden. Mögliche Ergebnisse des Peer Review reichen von der Bestätigung des eingeschlagenen Weges über bestimmte Modifikationen bis hin zur Empfehlung größerer Veränderungen im Verfahren.

Das StandAG selbst nimmt in dieser Hinsicht eine ambivalente Rolle ein. Einerseits legt es die Verfahrensschritte fest, ohne die die beteiligten Institutionen nicht agieren könnten. Dadurch ist das Gesetz einerseits notwendiger Hüter der Kontinuität des Verfahrens. Diese Kontinuität reicht intendiert bis in den Bereich der geowissenschaftlichen Ausschlusskriterien, der Mindestanforderungen und der Abwägungskriterien. Diese wurden im Gesetz festgeschrieben, um Konsistenz und Vergleichbarkeit aller Schritte im Verfahren zu gewährleisten. Denn eine Änderung von Kriterien könnte zu Inkonsistenzen führen und theoretisch sogar einen Neustart erfordern. Andererseits jedoch ist es angesichts der langen Zeitdauer des Verfahrens nicht unplausibel, dass neue wissenschaftliche Erkenntnisse – die z. B. in einem internationalen Peer Review zur Sprache gebracht werden könnten – Anlass geben, einige der Kriterien zu überdenken oder neue hinzuzunehmen. Dann stünden der Wunsch nach Kontinuität und Vergleichbarkeit auf der einen und der internationale Stand von Wissenschaft und Technik auf der anderen Seite in einem Widerspruch. Vermutlich müsste in diesem Fall das StandAG novelliert werden, mit möglicherweise weitreichenden Folgen. Hier wird das Prinzip des Lernens und der Selbsthinterfragung auf die Spitze getrieben: Das StandAG würde vom regulierenden Rahmen selbst zum Objekt der Veränderung. In einer solchen Situation käme dem zumindest weitgehenden demokratischen Konsens für die Novellierung eine entscheidende Bedeutung zu. Dieses Gedankenspiel macht deutlich, dass das StandAG nicht einfach „abgearbeitet" werden darf.

6 Schlussbemerkung

Das Monitoring ist fast schon ein geflügeltes Schlagwort geworden, ohne dass immer klar wird, aus welchen Gründen und zu welchen Zwecken es betrieben werden und was über das Monitoring genau erreicht werden soll. Die

Endlagerkommission selbst widmete diesem Thema wie gezeigt nennenswerte Aufmerksamkeit, welche heute stark vom NBG getragen wird. Daraus ergeben sich verschiedene Möglichkeiten für soziotechnische Gestaltungsprozesse bis hin auch zur Grundlagenforschung (Lösch 2021). Eine zentrale soziotechnische Frage dürfte dabei sein, wie wissenschaftlich-technisch motivierte Diagnosen und Empfehlungen durch Selbstreflexion der interessierten Öffentlichkeit aufgenommen und letztlich in Gestaltungsaufgaben umgemünzt werden.

Die Anforderungen an Lernen, Selbsthinterfragung und die resultierenden Erwartungen an das Monitoring umfassen freilich auch die Auseinandersetzung mit Perspektiven außerhalb des im Standortauswahlverfahren verfolgten Entsorgungspfades der Tiefenlagerung. Zumindest müssen die Nebenfolgen der nuklearen Entsorgung in einem Tiefenlager und der ihnen zugrunde liegenden Prozesse früh in den Blick genommen werden. Hier sind Einschätzungen vorzunehmen, wie Lasten räumlich und zeitlich verteilt werden und wie die Wirkungen dieser Verteilung soziale Räume und vorhandene Strukturen der Abfall- und Entsorgungspolitik fortschreiben. Die Frage der Konzentrierung der hochgefährlichen Abfälle an einem Ort oder ihrer dezentralen Konditionierung und Einlagerung wurde zwar von der Endlagerkommission zugunsten einer zentralen Lösung beantwortet, vom Deutschen Bundestag in das StandAG aufgenommen und wird zurzeit von den beteiligten Institutionen umgesetzt. Der Devise der Endlagerkommission nach muss aber selbst eine so weitreichende Festlegung revidierbar sein. Unter welchen Bedingungen die im Standortauswahlverfahren festgeschriebene Reversibilität ernsthaft für eine massive Umsteuerung ins Feld geführt werden wird, wird eventuell die Zukunft zeigen.

Literatur

EK – Kommission Lagerung hoch radioaktiver Abfallstoffe (2016): Verantwortung für die Zukunft – Ein faires und transparentes Verfahren für die Auswahl eines nationalen Endlagerstandortes. Abschlussbericht. https://www.bundestag.de/endlager/ [Zugriff am 14.09.2021]

FG IV – Fachgruppe IV des Nationalen Begleitgremiums (2021): Homepage mit Beschreibung seiner Mission und der Aktivitäten. https://www.nationales-begleitgremi-um.de/DE/WasWir Machen/Fachgruppen/_docs/texte/FG_4_Selbsthinterfragendes_Verfahren.html [Zugriff am 28.06.2021]

Grunwald, A. (2016): Der lange Weg zum Konsens. Abschlussbericht der Endlagerkommission. Politische Ökologie 146, S. 124–127

Hocke, P.; Bechthold, E.; Kuppler, S. (Hg.) (2016): Rückholung der Nuklearabfälle aus dem früheren Forschungsbergwerk Asse II. Dokumentation einer Vortragsreihe am Institut für Technikfolgenabschätzung und Systemanalyse (ITAS). Karlsruhe: KIT Scientific Working Papers Nr. 47; DOI(KIT): 10.5445/IR/1000063867

Hocke, P.; Bergmans, A.; Kuppler, S. (2012): Gewährleistung von Transparenz bei der Entsorgung nuklearer Abfälle: Monitoring als soziale Innovation. Einführung in den Schwerpunkt. Technikfolgenabschätzung – Theorie und Praxis 21(3), S. 5–10

Kuppler, S.; Hocke, P. (2019): The role of long-term planning in nuclear waste governance. Journal of risk research 22(11), S. 1343–1356

Lösch, A. (2021): Welche Unterscheidungen braucht die Endlagerforschung? Soziotechnische Gestaltung zwischen Möglichkeit und Unmöglichkeit. In: Brohmann, B.; Brunnengräber, A.; Hocke-Bergler, P.; Isidoro Losada, A.M. (Hg.): Robuste Langzeit-Governance bei der Endlagersuche. Soziotechnische Herausforderungen im Umgang mit hochradioaktiven Abfällen. Edition Politik 115, Bielefeld, S. 25–43

NBG – Nationales Begleitgremium (2021): Homepage mit Beschreibung seiner Mission und der Aktivitäten: https://www.nationales-begleitgremium.de/DE/Home/home_node.html [Zugriff am 28.06.2021]

Smeddinck, U. (2019): Sanfte Regulierung: Ressourcen der Konfliktlösung im Standortauswahlverfahren für ein Endlager. Deutsches Verwaltungsblatt (DVBl) 134(12), S. 744–751

StandAG (2017): Gesetz zur Suche und Auswahl eines Standortes für ein Endlager für Wärme entwickelnde radioaktive Abfälle (Standortauswahlgesetz – StandAG). Standortauswahlgesetz vom 5. Mai 2017 (BGBl. I S. 1074), das zuletzt durch Artikel 1 des Gesetzes vom 7. Dezember 2020 (BGBl. I S. 2760) geändert worden ist.

Melanie Mbah

Participation in decision-making processes as a key to a successful long-term governance[1]

1 Introduction

Recent developments in decision-making on contested infrastructure projects indicate a participatory turn, for example in nuclear waste governance (Bergmans et al. 2015; Seidl et al. 2013), but also in other fields such as renewable energy or transport infrastructure (Renn/Schweizer 2009; Römmele/Schober 2013). The reason for this is that traditional government approaches have lost their steering capacity in solving complex and wicked problems[2] (Brunnengräber et al. 2014).

Some authors therefore diagnose a current crisis of representative democracy (e.g., Kriesi 2013). This crisis is also referred to as post-democracy (Crouch 2004). The background of this diagnosis is the growing number of protests, citizens' groups, and social movements demanding more transparency, participation, and democratization (Dalton 2014; Geißel 2012), but also the restructuring of nationalism, for instance in Germany in the form of Pegida[3] or the AfD[4]. Therefore, new concepts of governance (e.g., Grande 2012; Hoppe 2010) put emphasis on the process and institutional structure as well as the design of decision-making.

---

1 This chapter is based on a contribution to the STS Conference Graz 2018 (Mbah 2018): Decision-Making in Repository Siting Procedures – Democratic and Societal Challenges for Nuclear Waste Governance. In: Getzinger, G. (ed.): Critical Issues in Science, Technology and Society Studies. Conference Proceedings of the 17th STS Conference Graz 2018, Graz, pp. 186–195.

2 Wicked problems are somehow unsolvable or unstructured. This indicates the need to organize learning processes that include a pluralism of perspectives (Brunnengräber et al. 2014; Hoppe 2010, pp. 228–229).

3 PEGIDA stands for "Patriotic Europeans against the Islamification of the Occident" and is a far-right nationalist movement.

4 AfD is the name of a new political party in Germany called Alternative for Germany, which represents right-wing nationalist views.

In nuclear waste governance, which is a complex social and technological problem (Hocke/Grunwald 2009), participation[5] is required to implement a robust decision-making process with the probable outcome of a decision that might gain acceptability[6]. Institutionalizing participation becomes even more important when considering the long time span from siting to closure of the repository. When it comes to long-term governance, participation should not stop after siting for several reasons. For example, many local/regional as well as national/global effects only become visible after a technology has been implemented (Kuppler/Mbah 2018). Instead, institutionalized participation in long-term governance could contribute to functioning checks and balances by ensuring continued public interest and vigilance (Kuppler 2017). Furthermore, there is an institutional need for strong interfaces with the public, because the interests of the general public and those of localities should be taken into account in a transparent, deliberative governance system in order to achieve acceptability, consider possible unwanted side-effects, and to keep decisions aligned with the common good (Kuppler/Mbah 2018; Kuppler/Hocke 2018, 2015). The term localities refers to place-specific interests of the future host community of the nuclear waste repository and the surrounding communities. A decision-making process that allows for the integration of different perspectives is more likely to be oriented toward the common good. Management literature shares the view that participatory processes are needed to integrate stakeholder knowledge into institutionalized decision-making processes (see, e.g., McLain/Lee 1996; Schindler/Cheek 1999). The complexity of the problem results from technical and social issues that are closely intertwined. Not only security and safety standards are important but also fairness – in terms of distribution, procedure, and outcome (Krütli et al. 2012).

In this paper I will analyze how participation is currently understood in nuclear waste governance and what challenges arise from the integration of more participatory elements in democratic decision-making – now and in the long run.

5 There is no common understanding of participation in the literature, therefore I base my argumentation on an understanding of participation as the active involvement of people affected or potentially affected as well as the interested public. Their interests are included in decision-making processes through participatory formats (see also Schaal/Ritzi 2012; van Deth 2009).

6 Acceptability means that legitimate democratic decisions are accepted although not all preferences and interests are fulfilled or satisfied (Grunwald 2005).

2 Participation in nuclear waste governance in Germany

In the past, repository siting processes in representative democracies often failed because decision-making processes were too focused on technical issues and decisions were made top-down without considering the concerns of the interested public (Hocke/Kallenbach-Herbert 2015; Krütli et al. 2010; Mackerron/Berkhout 2011). In the long run, this led to conflicts and hardened fronts between various citizens' groups and the broader public against government activities on all nuclear energy issues (Hocke/Kallenbach-Herbert 2015; Hocke/Renn 2011). After the severe accident in Fukushima (in March 2011), protests against nuclear energy revived and forced the government to act, resulting in the German decision to phase out nuclear energy in June 2011. This led to a window of opportunity in nuclear waste governance (Brunnengräber et al. 2014), which culminated in the StandAG[7] (2014, revised version 2017) and the establishment of the Commission for High-Level Waste Disposal[8] (2014–2016). The site selection process for a nuclear waste repository is an outstanding participation process in Germany in that there has been none in the past that reached so far into the future and is so challenging in terms of socio-technical complexity and wickedness. In this process, participation is often understood as an instrument to achieve predetermined subgoals. In other words, the objective is already clear, e.g., building a specific infrastructure and participation is only a means to achieve this objective, without openness to alternative proposals. Here, participation becomes an instrument of a new mode of decision-making.

This instrumentalization of participatory elements is often criticized by participation advocates (e.g., Dalton 2014; Di Nucci et al. 2017; Lehtonen 2010). They argue for the recognition of different understandings of participation and for a critical differentiation of different functions. Lehtonen (2010, pp. 179–181) has developed a very useful and comprehensive classification of the different functions of participation, namely the normative, the instrumental, and the substantive function. The normative function focuses on the process of participation and deliberation to guarantee legitimacy, for instance, in terms of equality of resources, openness to arguments, and equal representation of interests. The instrumental function focuses on the goals of different interest groups and their motivations in order to build legitimacy and trust in government, authorities, and public decisions. The substantive function focuses on

7 StandAG (Standortauswahlgesetz – Gesetz zur Suche und Auswahl eines Standortes für ein Endlager für hochradioaktive Abfälle) is the German Site Selection Act for a high-level waste repository.

8 The Commission for High-Level Waste Disposal developed criteria and recommendations for the site selection process and reviewed the Site Selection Act.

the outcome of participatory and deliberative processes, i.e., the integration of a wide range of values, knowledge, and discourses in order to open up new perspectives and achieve better solutions.

Furthermore, criteria of participation are needed to evaluate and categorize different standards of participation. To this end, Renn (2013) distills the fulfillment of a good standard of participation down to four basic criteria of participation processes in infrastructure projects: First, fairness must be guaranteed. That is, participants have equal rights and duties. Second, there must be a common basis of competence, which should be ensured by the accessibility of information and knowledge. Third, rules and techniques of conversation and scopes of action of participants should be introduced and maintained in order to gain legitimacy. And last but not least, efficiency should be achieved by incorporating participation outcomes into decision-making (Renn 2013, p. 80). Such processes require the willingness of individuals within public authorities to give up or share privileges, for instance, in terms of knowledge, information, and power. At the same time, individuals need to be open to new perspectives and arguments. Especially a long-term governance institution should be a so-called learning organization. One of the prerequisites for becoming such a learning organization is that the individuals organizing and steering such governance processes are open to change, in terms of public demands, the general social and political context in which a repository is to be installed, and technical progress. Due to this dynamic, the knowledge and information gap cannot be easily reduced or even closed. Power relations are not simply balanced either, because they are not only maintained by status and official positioning but also by rhetorical skills, prior knowledge, and an individual's ability to process information (see also Geißel 2009). Accordingly, comprehensibility of information is as important as accessibility of knowledge and information. The interested public and NGO members as well as directly or indirectly affected persons must be prepared to work together with stakeholders and public authorities in participatory processes. This requires – beside resources[9] – commitment and openness to other perspectives in order to be able to reconsider and enhance one's own arguments. In particular the fourth criterion suggested by Renn (2013) – efficiency – is difficult to meet in Germany due to the lack of legal regulations at national level for the incorporation of participation outcomes into decision-making. Deliberation in informal and formal participatory processes could be one option to take participation outcomes into account (see also Geißel 2012). This means that decisions are still made by legitimate

9 In order to guarantee equality it is also important to give actors a chance to debate on an equal footing. Here financial and time resources come into play. For further reading see Hoppe (2010) and Geißel (2012, 2009).

decision-makers and not by the population, as would be the case in a national referendum. Dryzek (2010, pp. 11 f.) characterizes such deliberative spaces as "accountable spaces" where individuals from authorities meet with individuals from the interested public to deliberate in symmetric inter-open spaces in order to achieve legitimacy of collective decision-making.

For too long, nuclear waste governance has focused on the instrumental function, which means that participation was, and sometimes still is, understood as a tool to gain trust and legitimacy (e.g., Andersson et al. 2003; Bergmans et al. 2015; Blowers/Sundqvist 2010). International recommendations, such as those of the IAEA (2017), still focus on the instrumental function of participation in recommending communication and consultation. This is criticized by, e.g., Blowers and Sundquist (2010, p. 153) as "technocratic framing." Trust in a fair process that considers all interests equally is not easily gained, and information and communication may not be the only or the right tools. A shift in perspective toward the normative and substantive functions of participation would therefore be a step forward in reducing the instrumental nature of participation in nuclear waste governance (Mbah 2017). Recent literature on the role of participation in nuclear waste governance for solving this socio-technical problem focuses on the process of decision-making, which means that the normative function of participation increases in importance (e.g., Krütli et al. 2012; Visschers/Siegrist 2012). This also applies to the way participation is discussed in the context of long-term governance, because here participation is indeed supposed to be an instrument to avoid neglect and delegation of responsibility, but at the same time it is suggested that participation is qualitatively necessary to ensure continuous decision-making in the future that is oriented to the common good (Kuppler/Mbah 2018).

The realization that the process matters a lot is also reflected in the new steps toward participatory decision-making in nuclear waste governance in Germany, such as the Commission for High-Level Waste Disposal and the Repository Site Selection Act (StandAG). The StandAG (2017) provides the statutory framework for the entire site selection process. Part two of the StandAG describes the framework and elements of participation: submission of statements and participation in discussions, the National Citizens' Oversight Committee (Nationales Begleitgremium – NBG), the Sub-Areas Conference, Regional Conferences, and the Council of the Regions Conference. The Federal Office for the Safety of Nuclear Waste Management (BASE)[10] is the supervisory and licensing authority and responsible for the participatory processes in the siting procedure. BASE aims to establish a modern organizational culture

10 The BASE has been renamed in the end of 2019/beginning of 2020. The former designation was Federal Office for the Safety of Nuclear Waste Management (BfE).

in which continuous learning processes, self-criticism, and a positive error culture should be part of daily work and managment (BASE 2020, 2019; see also Brohmann et al. 2021 and Mbah/Brohmann 2021). This is to ensure a siting procedure of the highest quality. At the same time, this authority should also govern the phases after site selection, i.e., the operation and monitoring phase, even though there is no concept for this to date. A long-term governance institution should actually have a positive error culture as well as the other characteristics mentioned above (e.g., Brohmann et al. 2021; Mbah/Brohmann 2021), but in reality the BASE builds on conventional governance experiences as it is part of the German authority structure and has recruited its staff partly from the Federal Office for Radiation Protection (Bundesamt für Strahlenschutz – BfS) and other agencies. Therefore, the BASE's general structure is still strongly influenced by the traditional governance framework, but at the same time it has to implement new governance ideas (see also BfE 2018).

A very important participatory element of the StandAG and thus also for the BASE is the NBG (StandAG, § 8) whose main task is to accompany the site selection process with special attention to participation in order to guarantee an open and participatory dialogue between all actors based on transparency. The overall aim is to gain trust in the process itself. Participatory elements of the new site selection process such as the NBG and the regional conferences have their role in being informed by the government in order to inform the interested public and to formulate their own statements in terms of a consultation process (StandAG 2017).

The focus of participation in the new German site selection procedure generally remains on information and consultation as well as transparency. This indicates that the StandAG focuses on formal correctness, and participation is only possible within a tight framework (Hocke/Smeddinck 2017). With regard to the activities of the NBG to date, efforts are being made to promote a transparent and participatory process. The NBG has therefore chosen various communication and publication channels to present findings, statements, and recommendations and to critically and constructively accompany the site selection process (cf. Flüeler 2021; NBG 2021a, 2021b, 2020; Trénel 2021). However, it has not yet been clarified how results of participatory elements are incorporated into the decision-making process. This is true not only for findings and recommendations of the NBG but also for various other participatory elements, especially various informal elements such as workshops on specific topics or with specific groups of actors, which produce findings and recommendations that are often not systematically analyzed and incorporated into further concepts, strategies, and decisions (cf. DAEF 2021). Suggestions have been made to communicate findings not only in the annual reports of the NBG but also directly to the government if certain findings are considered very important or urgent (empirical findings of participant observation by Mbah 2018), but so far government responses have been low (cf. Deutscher Bundestag 2021).

In summary, although in German nuclear waste governance, but also in other policy fields, a transparent and participatory process that means more than information is still a novelty, the need for a shift in the understanding of the functions of participation toward normative and substantive functions has already been recognized by policymakers and the interested public. There is a political will to take participation seriously. A set of criteria for a good standard of participation processes has been defined, but the implementation of these criteria is difficult in practice because it is influenced to a great extent by authorities and individual actions and their abilities and willingness to learn. Furthermore, questions of the compatibility of participatory elements and the underlying claim of participation as codetermination in decision-making with the representative democratic model are still open and need to be solved.

3 New challenges through participation

The traditional role of participation in representative democracy is quite limited. Participation is based on elections and membership in political parties or interest groups. In Germany, other forms of small-scale participation are possible at local and federal government levels, e.g., hearings, city hall meetings, and referenda. The fundamental characteristics of representative democracy are based on several principles (Biegelbauer/Hansen 2011). These include equality, representation, and accountability. Equality and representation are intertwined, which means that representation of all interests is the overall aim, irrespective of career or social status. Consequently, decision-makers should be representative of society, and decision-making processes should not be dependent on minority interests (Biegelbauer/Hansen 2011; Saward 2016).

According to the representative democracy model, the main challenge in implementing participatory elements is to ensure equality and representativity in decision-making processes (see Hoppe 2010; Michels 2011). Decker and Fleischer (2012) analyzed citizen dialogues on future technologies initiated by the Federal Ministry of Education and Research in 2010 and 2011 and discussed problems of equality and representativity in participatory processes as well as how to consider the outcomes of these dialogues in decision-making. They concluded that expectations of representativity were not fulfilled although each dialog involved more than 1,000 citizens selected by random sampling. In practice, the dialogues showed a bias toward older, highly educated male participants due to voluntary self-selection processes (Alcantara et al. 2016; Benighaus et al. 2016). Nevertheless, Decker and Fleischer (2012, p. 97) came to the conclusion that if the quality of deliberation could be met (1,000 representative citizens in table rounds of eight to ten people) "the outcomes could be considered as representative" and put "pressure on the government to actually implement the recommendations [...] and not just to consider it as 'another form of advice'" (ibid., p. 97).

Apart from that, the societal challenge is not only to ensure the efficiency of participatory elements in decision-making, but participation should be more than information and tend toward codetermination. Regarding equality and thus legitimacy of participatory processes, political authorities and decision-makers should keep in mind that power can be exercised to a greater extent in participatory processes than in representative decision-making processes (Geißel 2009). Moreover, it is difficult to identify all preferences of the people concerned because information costs are extremely high (Feindt/Newig 2005). As a consequence, not all affected individuals can be informed and participate in the process, vested interests may prevail, and participants might be overwhelmed by the complexity of the issues. This can lead to selectivity, which in turn challenges equality (see Craig 2014; Swyngedouw 2011, 2005; Thaa 2016).

In summary, there are different forms of participation and the literature reveals controversies about what participation means and when participation contradicts the existing model of representative democracy in Germany. In terms of German law, the implementation of elements of direct democracy, especially at national level, is critical. For example, referenda at national level are not compatible with the parliamentary system because the Federal Assembly would be undermined (Decker 2014; Linden 2016). Moreover, elements of direct democracy can be problematic in terms of legitimacy due to the selectivity of participants, which might have a negative effect on the equality of all interests. This argument is based on findings that participation in referenda is more unequal than in elections (Decker 2014; Schäfer/Schoen 2013). With regard to deliberative elements, it is easier to meet the criteria of legitimacy within the scope of representative decision-making. In other words, if a decision is still made in parliament, which means that deliberation outcomes are only recommendations, then deliberation can be evaluated as not competing with representative standards. Still, this form of participation is criticized as not being real participation, as the participants are only allowed to have a say, but not to decide. Participation according to the StandAG still moves within the framework of the representative democracy model (see also Hocke/Smeddinck 2017). However, it is an attempt to embed the informal participation procedures envisioned in the StandAG in a formal normative framework, i.e., the StandAG provides for the possibility to implement informal participation formats that are at the same time not mandatory if the minimum requirements of formal participation are fulfilled (Haug/Zeccola 2018, pp. 79 f.). In Germany, different ideas prevail on the type of participation desired. Actors highlight different challenges in the debate, such as who is responsible, how to ensure legitimacy, or how to prevent decision-making behind closed doors.

4 Conclusion

With regard to nuclear waste governance, there is a clear need for participation in the decision-making process, especially to fulfill the requirements for long-term governance of the waste. Based on this finding, ways must be found to integrate participation into decision-making that are sufficient and satisfactory to participants, stakeholders, and the interested public. It remains to be seen whether the StandAG serves this goal and these principles and whether the central actors – BASE, BGE, and the NBG – will be effective in facilitating an acceptable standard of participation. In the long term, these institutions will have to show whether they are able to adapt to changing contexts. This will only be possible if these organizations are organized as learning institutions that take into account the interests of the general public and the localities in their decision-making procedures. However, acceptability also means that participatory elements have to be in line with the representative democracy model and its decision-making structures.

In summary, decisions that are based on participatory processes are not necessarily better. For instance, in terms of content it might mean that not the geologically best site is chosen, but a site that is acceptable for other reasons, such as proximity to existing nuclear infrastructure, as has happened in Sweden and Finland (Forsmark/Sweden and Olkiluoto/Finland) (Kåberger/Swahn 2015; Lehtonen et al. 2017). Another example is that it might not be better in terms of representation because equal representation of all social groups is not ensured, but participants and the interested public might still value the participatory process in terms of transparency, inclusion of more perspectives, and enhancement of knowledge, making decisions more accountable and legitimate. Both examples also account for the phases after siting – the operational and monitoring phases – because a decision-making process that enables the integration of different views is more likely to be oriented toward the common good. Consequently, both procedural fairness and outcome fairness are important for the acceptability of decisions. Establishing procedural fairness might be easier than influencing attitudes toward nuclear energy and nuclear waste governance in order to achieve outcome fairness (Visschers/Siegrist 2012). Therefore, the first step should be the implementation of a fair process, which could then create acceptability. Regarding the wicked problem of nuclear waste storage and disposal, a broad and open debate as well as early involvement of stakeholders and the interested public might help to open up new perspectives. Participation needs to be directed toward codetermination in the sense of a higher level of participation, as described in Arnstein's "ladder of participation" (1969) as "partnership" or even "delegated power," to ensure a long-term public and governmental interest in governing nuclear waste even after closure of the repository. Reversibility and recurrent validation of possible alternatives should also be part of robust nuclear waste governance. The StandAG, with its new semi-institutional framework of participation, the NBG, is one step toward a more participatory approach, but more needs to be done to enable discourse on an equal footing and to integrate the results of participatory elements into decision-making. Therefore, a clear concept of the participatory process is needed, which is one of the main tasks of the BASE. This also includes the validation of alternatives and, of course, the question of how the findings of participatory processes will be integrated into decision-making on nuclear waste governance with regard to democratic structures and over a long period of time.

References

Alcantara, S.; Bach, N.; Kuhn, R.; Ullrich, P. (eds.) (2016): Demokratietheorie und Partizipationspraxis: Analyse und Anwendungspotentiale deliberativer Verfahren. Wiesbaden

Andersson, K.; Westerlind, M.; Atherton, E.; Besnus, F.; Chataîgnier, S.; Engström, S.; Espejo, R.; Hicks, T.; Hedberg, B.; Hunt, J.; Laciok, A.; Leskinen, A.; Lilja, C.; O'Donoghue, M.; Pierlot, S.; Wene, C.-O.; Vira, J.; Yearsley, R. (2003): Transparency and Public Participation in Radioactive Waste Management. RISCOM II Final Report; https://www.stralsakerhetsmyndigheten.se/contentassets/875908ece0364f338ecaa174c0031529/200408-transparency-and-public-participation-in-radioactive-waste-management-pdf-575-kb [Accessed 15.04.2018]

Arnstein, S.R. (1969): A Ladder of Citizen Participation. Journal of the American Institute of Planners 35(4), pp. 216–224

BASE – Bundesamt für die Sicherheit der nuklearen Entsorgung (2019): Information, Dialog, Mitgestaltung – Öffentlichkeitsbeteiligung in der Startphase der Endlagersuche. Anregungen aus dem Konsultationsprozess zum Konzeptentwurf. Berlin

BASE – Bundesamt für die Sicherheit der nuklearen Entsorgung (2020): Sicherheit der nuklearen Entsorgung – unsere Grundsätze. Berlin

Benighaus, C.; Wachinger, G.; Renn, O. (2016): Bürgerbeteiligung: Konzepte und Lösungswege für die Praxis. Frankfurt am Main

Bergmans, A.; Sundquist, G.; Kos, D.; Simmons, P. (2015): The Participatory Turn in Radioactive Waste Management: Deliberation and the Socio-Technical Divide. Journal of Risk Research 18(3), pp. 347–363

BfE – Bundesamt für kerntechnische Entsorgungssicherheit (2018): Unterschiedliche Rollen – ein Ziel. Positionspapier des BfE zur Öffentlichkeitsbeteiligung in der Standortauswahl. Berlin

Biegelbauer, P.; Hansen, J. (2011): Democratic Theory and Citizen Participation: Democracy Models in the Evaluation of Public Participation in Science and Technology. Science and Public Policy 38(8), pp. 589–597

Blowers, A.; Sundqvist, G. (2010): Radioactive Waste Management – Technocratic Dominance in an Age of Participation. Journal of Integrative Environmental Sciences 7(3), pp. 149–155

Brohmann, B.; Mbah, M.; Schütte, S.; Ewen, C.; Horelt, M.-A.; Hocke, P.; Enderle, S. (2021): Öffentlichkeitsbeteiligung bei der Endlagersuche: Herausforderungen eines generationenübergreifenden, selbsthinterfragenden und lernenden Verfahrens. Schlussfolgerungen und Empfehlungen. Projekt-Abschlussbericht im Auftrag des BASE, Vorhaben-Nr. 4717F00001; 04/2021; BASE: Berlin; urn:nbn:de:0221–2021051027029

Brunnengräber, A.; Di Nucci, M.R.; Häfner, D.; Isidoro Losada, A.M. (2014): Nuclear Waste Governance – ein *wicked problem* der Energiewende. In: Brunnengräber, A.; Di Nucci, M.R. (eds.): Im Hürdenlauf zur Energiewende. Von Transformationen, Reformen und Innovationen. Wiesbaden, pp. 389–399

Craig, T. (2014): Citizen Forums against Technocracy? The Challenge of Science to Democratic Decision Making. Perspectives on Political Science 43, pp. 31–40

Crouch, C. (2004): Post-democracy. Malden, MA

DAEF – Deutsche Arbeitsgemeinschaft Endlagerforschung (2021): "Lernendes Verfahren im Standortauswahlverfahren": Empfehlungen und Angebote der Deutschen Arbeitsgemeinschaft Endlagerforschung. Clausthal, Dresden; https://www.endlagerforschung.de/assets/daef_lernverf_thesenpapier_final_2021–05–12.pdf [Accessed 27.01.2022]

Dalton, R.J. (2014): Citizen Politics: Public Opinion and Political Parties in Advanced Industrial Democracies. Los Angeles

Decker, F. (2014): Volksgesetzgebung und parlamentarisches Regierungssystem: Eine schwierige Kombination. In: Münch, U.; Hornig, E.-C.; Kranenpohl, U. (eds.): Direkte Demokratie. Analysen im internationalen Vergleich. Baden-Baden, pp. 23–37

Decker, M.; Fleischer, T. (2012): Participation in "Big Style": First Observations at the German Citizens' Dialogue on Future Technologies. Poiesis & Praxis: International Journal of Ethics of Science and Technology Assessment 9(1/2), pp. 81–99

Deutscher Bundestag (2021): Nationales Begleitgremium fordert bessere Bürgerbeteiligung. Umwelt, Naturschutz und nukleare Sicherheit/Ausschuss, 23.06.2021, hib 831/2021; https://www.bundestag.de/presse/hib/849350-849350 [Accessed 27.01.2022]

Di Nucci, M.R.; Brunnengräber, A.; Isidoro Losada, A.M. (2017): From the "Right to Know" to the "Right to Object" and "Decide". A Comparative Perspective on Participation in Siting Procedures for High Level Radioactive Waste Repositories. Progress in Nuclear Energy 100, pp. 316–325

Dryzek, J.S. (2010): Foundations and Frontiers of Deliberative Governance. Oxford

Feindt, P.H.; Newig, J. (2005): Politische Ökonomie von Partizipation und Öffentlichkeitsbeteiligung im Nachhaltigkeitskontext: Probleme und Forschungsperspektiven. In: Feindt, P.H. (ed.): Partizipation, Öffentlichkeitsbeteiligung, Nachhaltigkeit. Perspektiven der politischen Ökonomie. Marburg, pp. 9–40

Flüeler, T. (2021): Öffentlichkeitsbeteiligung in der Beteiligungslücke nach Schritt 1 – aber sicher! Ein Blick von außen auf und Empfehlungen für Schritt 2 der Phase 1 des Standortauswahlverfahrens für ein Endlager für hochradioaktive Abfälle. Gutachten im Auftrag des Nationalen Begleitgremiums (NBG); https://www.nationales-begleitgremium.de/SharedDocs/Downloads/DE/Downloads_Gutachten/Gutachten_Verfahren_Flueeler_29_10_2021.html [Accessed 28.01.2022]

Geißel, B. (2012): Democratic Innovations: Theoretical and Empirical Challenges of Evaluation. In: Geißel, B.; Newton, K. (eds.): Evaluating Democratic Innovations. Curing the Democratic Malaise? New York, pp. 209–214

Geißel, B. (2009): Participatory Governance: Hope or Danger for Democracy? A Case Study of Local Agenda 21. Local Government Studies 35(4), pp. 401–414

Grande, E. (2012): Governance-Forschung in der Governance-Falle? Eine kritische Bestandsaufnahme. Politische Vierteljahresschrift (PVS) 53(4), pp. 565–592

Grunwald, A. (2005): Zur Rolle von Akzeptanz und Akzeptabilität von Technik bei der Bewältigung von Technikkonflikten. TATuP 14(3), pp. 54–60

Haug, V.M.; Zeccola, M. (2018): Neue Wege des Partizipationsrechts – eignet sich das Standortauswahlgesetz als Vorbild? ZUR 75, pp. 1–15

Hocke, P.; Grunwald, A. (eds.) (2009): Wohin mit dem radioaktiven Abfall? Berlin

Hocke, P.; Kallenbach-Herbert, B. (2015): Always the Same Old Story? Nuclear Waste Governance in Germany. In: Brunnengräber, A.; Mez, L.; Di Nucci, M.R.; Schreurs, M. (eds.): Nuclear Waste Governance. An International Comparison. Wiesbaden, pp. 177–201

Hocke, P.; Renn, O. (2011): Concerned Public and the Paralysis of Decision-Making: Nuclear Waste Management Policy in Germany. In: Strandberg, U.; Andren, M. (eds.): Nuclear Waste Management in a Globalised World, Abingdon, pp. 43–62

Hocke, P.; Smeddinck, U. (2017): Robust-parlamentarisch oder informell-partizipativ? Die Tücken der Entscheidungsfindung in komplexen Verfahren. GAIA – Ecological Perspectives for Science and Society 26(2), pp. 125–128

Hoppe, R. (2010): The Governance of Problems: Puzzling, Powering, Participation. Bristol

IAEA – International Atomic Energy Agency (2017): Communication and Consultation with Interested Parties by the Regulatory Body. General Safety Guide 6. Vienna; http://www-pub.iaea.org/MTCD/Publications/PDF/P1784_web.pdf [Accessed 03.04.2018]

Kåberger, T.; Swahn, J. (2015): Model or Muddle? Governance and Management of Radioactive Waste in Sweden. In: Brunnengräber, A.; Mez, L.; Di Nucci, M.R.; Schreurs, M. (eds.): Nuclear Waste Governance. An International Comparison. Wiesbaden, pp. 203–225

Kriesi, H. (2013): Democratic Legitimacy: Is There a Legitimacy Crisis in Contemporary Politics? Politische Vierteljahresschrift (PVS) 54(4), pp. 609–638

Krütli, P.; Stauffacher, M.; Flüeler, T.; Scholz, R.W. (2010): Functional-Dynamic Public Participation in Technological Decision-Making: Site Selection Processes of Nuclear Waste Repositories. Journal of Risk Research 13(7), pp. 861–875

Krütli, P.; Stauffacher, M.; Pedolin, D.; Moser, C.; Scholz, R.W. (2012): The Process Matters: Fairness in Repository Siting For Nuclear Waste. Social Justice Research 25(1), pp. 79–101

Kuppler, S. (2017): Effekte deliberativer Ereignisse in der Endlagerpolitik. Wiesbaden

Kuppler, S.; Hocke, P. (2018): The role of long-term planning in nuclear waste governance. Journal of Risk Research 21(3), pp. 1–14

Kuppler, S.; Hocke, P. (2015): "Enabling" public participation in a social conflict. The role of long-term planning in nuclear waste governance. In: Scherz, C.; Michalek, T; Hennen, L.; Hebáková, L.; Hahn, J; Seitz, S. (eds.): The next horizon of technology assessment. Proceedings from the PACITA 2015 Conference, Berlin, pp. 121–125

Kuppler, S.; Mbah, M. (2018): Governing energy landscapes: The need for a long-term, place-sensitive perspective, oral presentation at the UFZ EnergyDays 2018: Energy Landscapes of Today and Tomorrow, 24–25 September 2018, Leipzig

Lehtonen, M. (2010): Deliberative Decision-Making on Radioactive Waste Management in Finland, France and the UK: Influence of Mixed Forms of Deliberation in the Macro Discursive Context. Journal of Integrative Environmental Sciences 7(3), pp. 175–196

Lehtonen, M.; Kojo, M.; Litmanen, T. (2017): The Finnish Success Story in the Governance of a Megaproject: The (Minimal) Role of Socioeconomic Evaluation in the Final Disposal of Spent Nuclear Fuel. In: Lehtonen, M.; Joly, P.B.; Aparicio, L. (eds.): Socioeconomic Evaluation of Megaprojects. Dealing with Uncertainties. London, pp. 83–110

Linden, M. (2016): Beziehungsgleichheit als Anspruch und Problem politischer Partizipation. Zeitschrift für Politikwissenschaft 26, pp. 173–195

Mackerron, G.; Berkhout, F. (2011): Learning to Listen: Institutional Change and Legitimation in UK Radioactive Waste Policy. In: Strandberg, U.; Andren, M. (eds.): Nuclear Waste Management in a Globalised World, Abingdon, pp. 111–130

Mbah, M. (2018): Teilnehmende Beobachtung der Sitzung des Nationalen Begleitgremiums in Karlsruhe. English: Empirical Findings of the Participant Observation of the Meeting of the NBG in Karlsruhe, Karlsruhe

Mbah, M. (2017): Partizipation und Deliberation als Schlüsselkonzepte im Konflikt um die Endlagerung radioaktiver Abfälle? Herausforderungen für die repräsentative Demokratie. ITAS-ENTRIA-Arbeitsbericht, 2017–01

Mbah, M.; Brohmann, B. (2021): Das Lernen in Organisationen. Voraussetzung für Transformationsprozesse und Langzeit-Verfahren. In: Brohmann, B.; Brunnengräber, A.; Hocke, P.; Isidoro Losada, A.M. (eds.): Robuste Langzeit-Governance bei der Endlagersuche. Soziotechnische Herausforderungen im Umgang mit hochradioaktiven Abfällen. Bielefeld, pp. 387–412

McLain, R.J.; Lee, R.G. (1996): Adaptive management: Promises and pitfalls. Environmental Management 20, pp. 437–448

Michels, A. (2011): Innovations in Democratic Governance: How Does Citizen Participation Contribute to Better Democracy? International Review of Administrative Sciences 77(2), pp. 275–293

NBG – Nationales Begleitgremium (2021a): Breites Engagement bei der Standortsuche fördern. 3. Tätigkeitsbericht. Berlin; https://www.nationales-begleitgremium.de/SharedDocs/Downloads/DE/Downloads_Bericht_NBG/3_Taetigkeitsbericht_NBG.pdf;jsessionid=145F6DD8521965E605776D41AECBE038.intranet221?__blob=publicationFile&v=2 [Accessed 27.01.2022]

NBG – Nationales Begleitgremium (2021b): Empfehlungen des Nationalen Begleitgremiums (NBG) zur Öffentlichkeitsbeteiligung in der Endlagersuche. Berlin; https://www.bundestag.de/resource/blob/848014/15584d8cfefacbed5f8539a9d71b1ae3/Nationales-Begleitgremium-NBG--data.pdf [Accessed 27.01.2022]

NBG – Nationales Begleitgremium (2020): Ein neuer Weg hat sich bewährt. Unsere Begleitung des Standortauswahlverfahrens. Berlin

Renn, O. (2013): Partizipation bei öffentlichen Planungen. Möglichkeiten, Grenzen, Reformbedarf. In: Keil, S.I.; Thaidigsmann, S.I. (eds.): Zivile Bürgergesellschaft und Demokratie. Wiesbaden, pp. 71–96

Renn, O.; Schweizer, P.J. (2009): Inclusive Risk Governance: Concepts and Application to Environmental Policy Making. Environmental Policy and Governance 19, pp. 174–185

Römmele, A.; Schober, H. (eds.) (2013): The Governance of Large-Scale Projects: Linking Citizens and the State. Baden-Baden

Saward, M. (2016): Fragments of Equality in Representative Politics. Critical Review of International Social and Political Philosophy 19(3), pp. 245–262

Schaal, G.S.; Ritzi, C. (2012): Deliberative Partizipation: Eine kritische Analyse des Verhältnisses von Deliberation, demokratischer Öffentlichkeit und staatlicher Entscheidung. In: Riescher, G.; Rosenzweig, B. (eds.): Partizipation und Staatlichkeit. Ideengeschichtliche und aktuelle Theoriediskurse. Stuttgart, pp. 131–153

Schäfer, A.; Schoen, H. (2013): Mehr Demokratie, aber nur für wenige? Der Zielkonflikt zwischen mehr Beteiligung und politischer Gleichheit. Leviathan 41(1), pp. 94–120

Seidl, R.; Krütli, P.; Moser, C.; Stauffacher, M. (2013): Values in the Siting of Contested Infrastructure: The Case of Repositories for Nuclear Waste. Journal of Integrative Environmental Sciences 10(2), pp. 107–125

Shindler, B.; Cheek, K.A. (1999): Integrating citizens in adaptive management: A propositional analysis. Conservation Ecology 3(1), article 9; https://www.researchgate.net/publication/42763224_Integrating_Citizens_in_Adaptive_Management_A_Propositional_Analysis [Accessed 02.11.2021]

StandAG (2017): Gesetz zur Suche und Auswahl eines Standortes für ein Endlager für Wärme entwickelnde radioaktive Abfälle (Standortauswahlgesetz – StandAG). Standortauswahlgesetz vom 5. Mai 2017 (BGBl. I S. 1074), das zuletzt durch Artikel 1 des Gesetzes vom 7. Dezember 2020 (BGBl. I S. 2760) geändert worden ist.

Swyngedouw, E. (2011): Interrogating Post-Democratization: Reclaiming Egalitarian Political Spaces. Political Geography 30(7), pp. 370–380

Swyngedouw, E. (2005): Governance Innovation and the Citizen: The Janus Face of Governance-Beyond-The-State. Urban Studies 42(11), pp. 1–16

Thaa, W. (2016): Issues and Images – New Sources of Inequality in Current Representative Democracy. Critical Review of International Social and Political Philosophy 19(3), pp. 357–375

Trénel, M. (2021): NBG-Gutachten zu der Auswirkung der Digital-Formate auf die Beteiligungsqualität der Fachkonferenz Teilgebiete. Gutachten im Auftrag des Nationalen Begleitgremiums (NBG); https://www.nationales-begleitgremium.de/SharedDocs/Downloads/DE/Downloads_Gutachten/Gutachten_Oeff-beteiligung_Trenel_25_10_2021.html [Accessed 28.01.2022] van Deth, J.W. (2009): Politische Partizipation. In: Kaina, V. (ed.): Politische Soziologie. Ein Studienbuch. Wiesbaden, pp. 141–161

Visschers, V.H.M.; Siegrist, M. (2012): Fair Play in Energy Policy Decisions: Procedural Fairness, Outcome Fairness and Acceptance of the Decision to Rebuild Nuclear Power Plants. Energy Policy 46, pp. 292–300

Ulrich Smeddinck, Franziska Semper

Long-term Governance zur Begleitung eines Endlagers aus rechtswissenschaftlicher Sicht

Beispiellos ist im Aufgabenkatalog der Bundesrepublik Deutschland, ein sicheres Endlager für hochradioaktive Abfälle (umgangssprachlich Atommüll[1]) für einen kaum vorstellbaren Zeitraum von einer Million Jahre zu realisieren. Beispiellos ist auch die Herausforderung für Recht und Administration, ein Endlager möglichst lange zu begleiten. Das bedeutet, ebenso Schwäche und Vergänglichkeit menschlicher Zivilisationen anzuerkennen (Breuer 1992, S. 174; Willke 2016, S. 264 f.) wie auch kreative Anpassungs- und Erneuerungsfähigkeit für möglich zu halten (Kromp/Lahodynsky 2006, S. 73; vgl. auch Leggewie/Welzer 2010, S. 174 ff.). Wie lässt sich eine solche Long-term Governance aus rechtswissenschaftlicher Sicht realisieren?

Der Beitrag verdeutlicht zunächst den Governance-Begriff in der Rechtswissenschaft (1), schildert bestehende Rechtsgrundlagen, die eine langfristige Perspektive einnehmen (2), und entwickelt erste Überlegungen zu Organisation und Verfahren unter Berücksichtigung zukünftiger Generationen im Hinblick auf eine langfristige institutionelle Begleitung eines Endlagers (3, 4).[2] In der Darstellung werden die Szenarien einer Long-term Governance unter den Bedingungen stabiler und sich auflösender Staatlichkeit unterschieden. Es wird auf weitere Forschungsperspektiven verwiesen (5). Der Beitrag schließt mit Fazit und Ausblick (6).

Da das Gesetz zur Suche und Auswahl eines Standortes für ein Endlager für hochradioaktive Abfälle (Standortauswahlgesetz)[3] über die enorme Zeitperspektive hinaus bisher keine konkreten Anknüpfungspunkte für Long-term Governance bietet, kann der Text im Weiteren auch keinen Bezug darauf nehmen, geht aber auf andere grundsätzliche und/oder sachdienliche Regelungen des Gesetzes ein.

---

1 Eine sprachkritische Auseinandersetzung mit dem Begriff findet sich bei Brunnengräber/Smeddinck (2016, S. 68 f.).

2 Vorüberlegungen bei Kloepfer (1993, S. 36 ff).

3 Vom 5. Mai 2017 (BGBl. I S. 1074), zuletzt geändert durch Gesetz vom 7. Dezember 2020 (BGBl. I S. 2760).

1 Governance-Begriff in der Rechtswissenschaft

Trotz erheblicher Aufmerksamkeit und Befassung mit dem Governance-Begriff ist dieser (jedenfalls in der Rechtswissenschaft) nicht abschließend geklärt (Voßkuhle 2012, § 1 Rz. 68 und 70). Er wird als Chiffre für den Modus und die Qualität modernen Regierens in komplexen Strukturen aufgefasst. Als erste Konkretisierung wird auf das Governance-Verständnis im Weißbuch Europäisches Regieren verwiesen: Governance wird dabei verstanden als das Ausformen von Regeln, Verfahren und Verhaltensweisen, die die Art und Weise, wie auf europäischer Ebene Befugnisse ausgeübt werden, kennzeichnen, und zwar insbesondere in Bezug auf Offenheit, Partizipation, Verantwortlichkeit, Effektivität und Kohärenz (Voßkuhle 2012, § 1 Rz. 68). Entsprechend steht Governance aus rechtswissenschaftlicher Sicht für die Abkehr von staatszentrierten Modellvorstellungen, bei denen der Staat als Steuerungssubjekt die Steuerungsobjekte steuert, und für die Förderung der Interaktion in und zwischen Netzwerken (Schuppert 2007, S. 463 ff.; Blumenthal 2005, S. 1163 ff.) – unter grundsätzlichem Festhalten am Steuerungsbegriff (Voßkuhle 2012, § 1 Rz. 70; vgl. auch Reimer 2016, S. 50).

2 Bestehende Rechtsgrundlagen mit langfristiger Perspektive

Es sind bereits eine ganze Reihe von Rechtsgrundlagen, die die Entsorgung des Atommülls in den Blick nehmen, in Kraft, die in langfristiger Perspektive die damit verbundenen Aufgaben und Probleme adressieren (vgl. auch Lersow/Gellermann 2015, S. 178 f.). Ihre Anwendung und Einhaltung geht unausgesprochen von der Grundüberzeugung aus, dass Long-term Governance unter den Bedingungen stabiler Staatlichkeit realisiert werden kann.

2.1 Langfristigkeit im Grundgesetz

Zwei Regelungen des Grundgesetzes (GG) betonen die Langfristigkeit in besonderer Weise:

In Art. 79 Abs. 3 findet sich die sog. Ewigkeitsklausel. Danach ist eine Änderung dieses Grundgesetzes unzulässig, durch welche die Gliederung des Bundes in Länder, die grundsätzliche Mitwirkung der Länder bei der Gesetzgebung oder die in den Artikeln 1 und 20 niedergelegten Grundsätze berührt werden. Damit wird unterstrichen, dass die gegenwärtige staatliche und verfassungsrechtliche Ordnung Geltung für immer beansprucht. Ansonsten würde „die Beseitigung des Grundgesetzes heraufbeschworen“ (Sachs 2011, Art. 79

Rz. 26). Die Grundsätze sollen im Übrigen auch vor einem allmählichen Zerfallsprozess geschützt werden (Sachs 2011, Art. 79 Rz. 37).

Dass sich disruptive Entwicklungen nicht verhindern lassen, hat die Geschichte ebenso gezeigt wie auch, dass die Feststellung der Unzulässigkeit bestimmter Aktivitäten durch Gerichte eine staatliche Ordnung nicht mehr retten kann, wenn sie denn ins Rutschen geraten ist (Friesenhahn 1980, S. 102). Auf unabsehbare Zeit kann zunächst aber von der Kontinuität der gegenwärtigen Ordnung ausgegangen werden.

Von der Ewigkeitsklausel nicht erfasst wird dagegen Artikel 20a GG, der einen eigenen zukunftsgerichteten Anspruch formuliert: Danach schützt der Staat auch in Verantwortung für die künftigen Generationen die natürlichen Lebensgrundlagen und die Tiere im Rahmen der verfassungsmäßigen Ordnung. Dies geschieht durch die Gesetzgebung und nach Maßgabe von Gesetz und Recht durch die vollziehende Gewalt und die Rechtsprechung. Die Vorschrift stellt eine Staatszielbestimmung dar. Es handelt sich um einen Optimierungsauftrag für die genannten staatlichen Akteure, dessen Einlösung von den Bürgerinnen und Bürgern nicht eingeklagt werden kann (Kluth 2016, Art. 20a Rz. 72).

2.2 Richtlinie 2011/70/Euratom

Von zentraler Bedeutung für die dauerhafte Verwahrung von Atommüll in Deutschland und Europa ist die Richtlinie 2011/70/Euratom des Rates vom 19. Juli 2011 über einen Gemeinschaftsrahmen für die verantwortungsvolle und sichere Entsorgung abgebrannter Brennelemente und radioaktiver Abfälle. Erwägungsgrund 21 Satz 1 hält fest, dass radioaktive Abfälle, einschließlich abgebrannter Brennelemente, die als Abfall angesehen werden, eingeschlossen und *langfristig* (Hervorhebung von US/FS) vom Menschen und der belebten Umwelt isoliert werden müssen. Erwägungsgründe, die dem eigentlichen Richtlinientext vorangestellt sind, entfalten allerdings keine verbindliche Regelungswirkung. Sie dienen lediglich als „Auslegungshilfe“ (Dietrich et al. 2003, S. 60.), in diesem Fall für die Richtlinie. Ebenso wie Zielbestimmungen in vielen Umweltgesetzen dienen sie neben den juristischen Auslegungsregeln (z. B. Smeddinck 2013, S. 21.) dazu, den eigentlichen Sinn unbestimmter Rechtsbegriffe zu ermitteln, aus denen Rechtsvorschriften typischerweise bestehen (Kluth 2014, § 1 Rz. 105 ff.).

Der Regelungszweck wird im rechtsverbindlichen Teil der Richtlinie in Artikel 1 Absatz 1 bestimmt. Dort heißt es: Mit dieser Richtlinie wird ein Gemeinschaftsrahmen für die verantwortungsvolle und sichere Entsorgung abgebrannter Brennelemente und radioaktiver Abfälle geschaffen, um zu vermeiden, dass künftigen Generationen unangemessene Lasten aufgebürdet werden

(Roßegger 2011, 276 ff.). Die Richtlinie knüpft einerseits an die Entsorgung abgebrannter Brennelemente an, die bei zivilen Tätigkeiten anfallen, und andererseits an die Entsorgung radioaktiver Abfälle, die bei zivilen Tätigkeiten anfallen, von der Erzeugung bis zur Endlagerung, und gilt für alle Stufen des Geschehens (Artikel 2 Absatz 1). Allerdings sind auch solche gesetzlichen Zielbestimmungen aus sich heraus nicht operationalisierbar (Kluth 2014, § 1 Rz. 106).

Echte Wirkung entfalten dagegen die Verpflichtungen in der Richtlinie. So müssen die Mitgliedstaaten einen nationalen Gesetzes-, Vollzugs- und Organisationsrahmen (im Folgenden „nationaler Rahmen") für die Entsorgung abgebrannter Brennelemente und radioaktiver Abfälle schaffen, der die Zuweisung der Verantwortlichkeit regelt und für die Koordinierung zwischen den einschlägigen zuständigen Stellen sorgt. Der nationale Rahmen muss u. a. ein nationales Programm zur Umsetzung der Politik für die Entsorgung abgebrannter Brennelemente und radioaktiver Abfälle enthalten. Konkretisierend heißt es dazu in Artikel 11 Absatz 1 Satz 1: Die Mitgliedstaaten stellen sicher, dass ihre nationalen Programme für die Entsorgung abgebrannter Brennelemente und radioaktiver Abfälle (im Folgenden „nationale Programme") durchgeführt werden und für Arten abgebrannter Brennelemente und radioaktiver Abfälle unter ihrer Rechtshoheit sowie alle Stufen der Entsorgung abgebrannter Brennelemente und radioaktiver Abfälle von der Erzeugung bis zur Endlagerung abdecken. Weiter verlangt die Richtlinie, dass die Mitgliedstaaten ihre nationalen Programme regelmäßig überprüfen und aktualisieren, wobei sie gegebenenfalls dem wissenschaftlichen und technischen Fortschritt sowie Empfehlungen, Erfahrungen und bewährten Praktiken, die sich aus den Prüfungen durch Experten ergeben, Rechnung tragen.

Gemäß Artikel 12 Absatz 1 e) gehören zum Inhalt der Programme Konzepte oder Pläne für den Zeitraum nach dem Verschluss innerhalb der Lebenszeit der Anlage zur Endlagerung, einschließlich des Zeitraums, in dem geeignete Kontrollen beibehalten werden, sowie der vorgesehenen Maßnahmen, um das Wissen über die Anlage längerfristig zu bewahren.

In Deutschland wird die Richtlinie insbesondere durch das *Standortauswahlgesetz* umgesetzt. In dessen § 1 Absatz 2 Satz 1 heißt es: Mit dem Standortauswahlverfahren soll in einem partizipativen, wissenschaftsbasierten, transparenten, selbsthinterfragenden und lernenden Verfahren für die im Inland verursachten hochradioaktiven Abfälle ein Standort mit der bestmöglichen Sicherheit für eine Anlage zur Endlagerung [...] in der Bundesrepublik Deutschland ermittelt werden. Der Standort mit der bestmöglichen Sicherheit ist der Standort, der im Zuge eines vergleichenden Verfahrens aus den in der jeweiligen Phase nach den hierfür maßgeblichen Anforderungen dieses Gesetzes geeigneten Standorten bestimmt wird und die bestmögliche Sicherheit für den dauerhaften Schutz von Mensch und Umwelt vor ionisierender Strahlung und sonstigen schädlichen Wirkungen dieser Abfälle für einen Zeitraum von einer Million Jahren gewährleistet (Absatz 2 Satz 2) (eingehend in Smeddinck 2016, § 1). Das Standortauswahlgesetz kreiert neue Akteure, die auf Dauer die Realisierung eines Endlagers begleiten sollen. Im Einzelnen sind das das Bundesamt für die Sicherheit der nuklearen Entsorgung (BASE), das als Kontrollbehörde fungieren soll (§ 4), der Vorhabenträger (die Bundesgesellschaft für Endlagerung – BGE), der die Standortsuche und das Endlager realisieren soll (§ 3), sowie das Nationale Begleitgremium (§ 8), dessen zentrale Aufgaben die vermittelnde und unabhängige Begleitung des Standortauswahlverfahrens, insbesondere auch der Umsetzung der Öffentlichkeitsbeteiligung am Standortauswahlverfahren bis zur Standortentscheidung nach § 20, sind (Schreurs 2017, S. 6 f.).

Resümierend lässt sich festhalten, dass die Langfristperspektive im bestehenden Recht und seine Fortschreibung bedacht und im üblichen Rahmen rechtsstaatlicher Administrierung gesichert sind. Fraglich ist aber, wie Long-term Governance unter der Bedingung einer sich auflösenden Staatlichkeit gewährleistet werden kann. Dabei richtet sich das Interesse nicht auf Entwicklungen innerhalb einer staatlichen Ordnung, die üblicherweise periodisch zu einer Umorganisation und Neuverteilung von Aufgaben zwischen Staat und Gesellschaft führt, wie das etwa die Ablösung des Wohlfahrtsstaates durch eher deregulierende, neoliberal-inspirierte Formen der Aufgabenerledigung illustriert. Gedacht ist eher an tatsächliche disruptive Entwicklungen, die die Funktionsfähigkeit bisheriger staatlicher Ordnungen zerstören oder auflösen (vgl. Diamond 2005, z. B. S. 628, 633, 644).

3 Verfahren unter Berücksichtigung zukünftiger Generationen

Nach der verfassten Staatlichkeit sollen nun die Eigeninteressen künftiger Generationen in den Blick genommen werden. Mit Erwägungsgrund 24 der Richtlinie 2011/70/Euratom wird bereits auf europäischer Ebene betont, dass es jedem Mitgliedstaat eine ethische Pflicht sein sollte, künftige Generationen in Bezug auf radioaktive Abfälle nicht unangemessen zu belasten. Aus diesem Grund sind auch die Interessen künftiger Generationen ein wichtiger Faktor, der bei den Erwägungen zur Long-term Governance berücksichtigt werden muss. Ferner „braucht es für rechtsverbindliche Entscheidungen klare Verfahren“ (Höffe 2015, S. 46). Zu den Fragen der Long-term Governance gehört zugleich die verfahrensrechtliche Ausgestaltung. Governance als Inbegriff von sowohl „formalen als auch informalen Steuerungsstrukturen“ (Reimer 2016, S. 51) ist am Gemeinwohl ausgerichtet. Neben dem Wohl der jetzigen Generationen gehört zum Gemeinwohl auch das Wohl der künftigen Generationen (Reimer 2016, S. 54). Da das Interesse der zukünftigen Generationen trotz hoher Relevanz oftmals vernachlässigt wird, soll im Weiteren der Fokus auf der Einbeziehung von künftigen Generationen in die Long-term Governance liegen.

3.1 Begründung der Einbeziehung zukünftiger Generationen – Prinzip des normativen Individualismus

Die zeitliche Zukunftsverantwortung umschließt die für uns heute überblickbare Zukunft, welche durch die Grenzen des prognostischen Wissens limitiert wird (Birnbacher/Schicha 2001, S. 20). Das bedeutet, dass die Vorsorge nicht weiter als zwei nachfolgende Generationen reichen kann. Diese Position vertritt auch Spaemann. Als Begründung führt er an, dass wir nicht wissen können, wie ferne Generationen leben werden. Rücksicht müssen wir nur auf „die Generationen der nächsten Jahrhunderte nehmen, die von den Ressourcen vielleicht besseren Gebrauch machen können“ (Spaemann 2003, S. 33; Hocke 2013, S. 266). Die Spanne der Zukunftsverantwortung wird im Verfahren der Entsorgung hochradioaktiver Reststoffe um mehr als zwei Generationen erweitert. Zum einen soll der Standort für ein Endlager die bestmögliche Sicherheit, für den dauerhaften Schutz für einen Zeitraum von einer Million Jahren, gewährleisten. Zum anderen muss nach § 1 Absatz 4 Satz 2 Standortauswahlgesetz die technische Möglichkeit bestehen, innerhalb eines Zeitrahmens von 500 Jahren nach dem geplanten Verschluss des Endlagers im Falle eines Notfalls die Bergung der Reststoffe vornehmen zu können (eingehend: Smeddinck 2021). Der langfristige Verfahrensprozess reicht in seiner Gesamtheit also über die übliche Zukunftsverantwortung hinaus. Eine besondere Berücksichtigung der Rechte und Interessen von zukünftigen Generationen erscheint notwendig.

Das Prinzip des normativen Individualismus besagt, dass „Pflichten und Werte in letzter Instanz nur durch Rekurs auf alle betroffenen Individuen und ihre Eigenschaften gerechtfertigt werden können“ (Pfordten 2014, S. 5). Damit bildet der normative Individualismus die ethische Grundlage für die Einbeziehung der Individuen zukünftiger Generationen, so unsere These. Bezugspunkt ist nicht das Kollektiv einer Generation, sondern das Individuum (Strack 2015, S. 250). Die letzte Rechtfertigung von Normen und Handlungen muss sich auf alle betroffenen Individuen stützen (Strack 2015, S. 251). Daraus folgt, dass im Falle der Betroffenheit zukünftiger Generationen, diese mit einbezogen werden müssen.

Der Begriff der Betroffenheit[4], der hier aufgegriffen wird, ist im rechtlichen Kontext vor allem im Rahmen des Verfassungsprozessrechts relevant. Von Betroffenheit wird gesprochen, wenn der Beschwerdeführer einer Verfassungsbeschwerde durch die Maßnahme der öffentlichen Gewalt beschwert ist. Die Maßnahme der öffentlichen Gewalt muss eine irgendwie geartete beeinträchtigende Einwirkung auf den persönlichen Schutzbereich eines Grundrechts des Beschwerdeführers haben (Bethge 2016, § 90 Rz. 351). Das Beispiel des Radionuklids[5] Plutonium 239 mit einer Halbwertszeit von ca. 24.000 Jahren verdeutlicht, dass bei der Endlagerung von hochradioaktiven Reststoffen unabhängig von der jeweiligen Entsorgungsoption zukünftige Generationen unmittelbar betroffen sind.

3.2 Rechte und Interessen der zukünftigen Generationen

Erweitert man nun unter Berücksichtigung der intergenerationellen Gerechtigkeit den Kreis der Beteiligten im Verfahrensablauf des Endlagerprozesses, stellt sich unter anderem die Frage, wie zukünftige Generationen in diesem Verfahren berücksichtigt werden können.

Zunächst ist zu klären, ob sie überhaupt Rechte haben. Recht als solches unterteilt sich im objektiven Sinne als die Gesamtheit der Rechtsordnung und im subjektiven Sinne als die Befugnis, die sich für den Berechtigten aus dem objektiven Recht herleitet (Creifelds 2014, S. 1019). Ob den zukünftigen Generationen subjektive Rechte zugesprochen werden können, ist umfassend diskutiert worden (Kloepfer 1993, S. 26 ff.; Tremmel 2012, S. 349 ff.; Ekardt 2010, S. 91). Bereits *Hans Jonas* betonte, „das Nichtexistierende stellt keine Ansprüche, kann daher auch nicht in seinen Rechten verletzt werden" (Jonas 1979, S. 84). Die überzeugende herrschende Ansicht geht davon aus, dass subjektive Rechte eine individualisierbare Existenz zwingend voraussetzen, um die kraft öffentlichen Rechts zuerkannte Rechtsmacht durchzusetzen (Kahl 2016a, S. 302). Es bedarf somit einer Rechtsträgerschaft, die zwingend an eine natürliche oder juristische Person gebunden ist. Aufgrund der Nichtexistenz können künftigen Generationen keine Rechte zugeschrieben werden. Sprachlich präziser sei es, dass künftige Generationen Rechte haben werden, sobald sie geboren sind (Tremmel 2012, S. 354). Es fehlt an der Person, die sich Gehör verschafft und gleichzeitig für ihre Rechte einsteht. Erst wenn die zukünftige Generation

---

4 Vgl. auch die frühe Öffentlichkeitsbeteiligung gemäß § 25 Abs. 3 des Verwaltungsverfahrensgesetzes (Kallerhoff 2013, § 25 Rz. 68: Adressat der frühen Öffentlichkeitsbeteiligung ist die betroffene Öffentlichkeit, die alle Personen, deren Belange durch das geplante Vorhaben oder das anschließende Verwaltungsverfahren berührt werden können, umfasst.

5 Die Kerne radioaktiver Atome heißen Radionuklide (vgl. Volkmer 2005, S. 8).

zur gegenwärtigen Generation wird, kann sie ihre Rechte im juristischen Sinne geltend machen. Aus diesem Grund kann die heutige Generation nur vermuten, welchen Willen oder welche Interessen die zukünftigen Generationen verfolgen werden.

Betrachtet man die künftigen Generationen nicht als Individuen, sondern nur als Kollektiv, dann könnten dem Kollektiv eventuell subjektive Rechte zustehen. Dem Kollektiv fehlt es allerdings an der notwendigen Individualisierbarkeit. Um Rechte geltend zu machen, braucht es einen potenziellen Rechtsträger (Kahl 2016a, S. 304).

Gleichwohl lassen sich rechtliche Schutzpflichten gegenüber den zukünftigen Generationen anhand von verschiedenen Verfassungsnormen begründen (Kahl 2016a, S. 306). Neben Artikel 1 Absatz 2 und Artikel 6 Absatz 1 GG ist es gemäß Artikel 20a GG die Aufgabe des Staates in Verantwortung für die künftigen Generationen die natürlichen Lebensgrundlagen und die Tiere zu schützen. Hier wird auf die zukunftsbezogenen Schutzpflichten des Staates Bezug genommen. Damit orientiert sich Artikel 20a GG an dem Leitgedanken der intergenerationellen Gerechtigkeit (Kluth 2016, Art. 20a Rz. 89).

Trotz fehlender subjektiver Rechte der künftigen Generationen erschließt sich eine Notwendigkeit der Einbeziehung von Belangen künftiger Generationen anhand der verfassungsrechtlichen Schutzpflichten. Diese zeigen, dass sie bis in die Zukunft reichen, um die erst zukünftig Lebenden als die künftige Bevölkerung in ihrer Existenz, Integrität und Freiheit zu bewahren (Kahl 2016a, S. 307).

Allein die materiell-rechtlichen Schutzpflichten führen aber nicht zu einer institutionellen oder verfahrensrechtlichen Berücksichtigung (Kluth 2016, Art. 20a Rz. 90). Somit müssen Möglichkeiten geschaffen werden, dass die Interessen von künftigen Generationen im Wege der Long-term Governance berücksichtigt werden.

3.3 Verfahrensbeteiligung

Ein Weg, diese Herausforderung zu bewältigen, ist, die Verpflichtungen gegenüber den nahestehenden Generationen allgemein mit einem höheren Grad an Verbindlichkeit zu organisieren. Rechtspflichten müssen präzisiert und verbindlich gemacht werden. Entsprechend kann der Klimabeschluss des Bundesverfassungsgerichts von 2021 gedeutet werden.[6] Eine konkrete Verpflichtung zur Handlung besteht so lange, wie mögliche Folgen zu Konfliktpotenzial führen können. Die Entsorgung von radioaktiven Reststoffen besteht für die Dauer

6 BVerfG, Beschluss vom 24. März 2021, 1 BvR 2656/18, 1 BvR 96/20, 1 BvR 78/20, 1 BvR 288/20, 1 BvR 96/20, 1 BvR 78/20.

der Zeit, bis von den radioaktiven Restoffen keine Gefährdung für Mensch und Umwelt mehr ausgeht (Streffer et al. 2011, S. 23). Aus diesem Grund erscheint es notwendig, im Zusammenspiel mit den Entscheidungsträgern die Interessen der zukünftigen Generationen in die Entscheidungsprozesse zu integrieren. Unter einer Integration im Verfahren ist vor allem die Einbeziehung unterschiedlicher Verfahrensschritte in ein Verwaltungsverfahren zu verstehen (Durner 2016, S. 324). Das Verfahren folgt einer wert- und erkenntnisorientierten Vorgehensweise. Im Vordergrund stehen die Unparteilichkeit und die gegenseitige Fairness. Integration heißt dann im Weiteren, dass möglichst alle entscheidungserheblichen Gesichtspunkte zur Geltung kommen, die Bedürfnisse und Interessen der Betroffenen offen einbezogen werden und die Bereitschaft zur Korrektur von Fehlern besteht. Geschieht dies nicht, so könnte eine einseitige Interessenvertretung der gegenwärtigen Generation dazu führen, dass das Entscheidungsverhalten der Verantwortlichen einseitig verzerrt wird (Streffer et al. 2011, S. 24). Um dies zu vermeiden, muss das Verfahren klar und nachvollziehbar gestaltet werden. Zuständigkeiten, Abläufe und Form dienen dem Inhalt und sind somit unabhängig vom Ergebnis (Höffe 2015, S. 46).

Eine weitere notwendige Voraussetzung sind die partizipativen Elemente in den jeweiligen Verfahrensschritten. Hilfreich könnte dabei ein neues Verfahrenselement sein. Die sogenannte „Nachweltverträglichkeitsprüfung“ ist bereits im Zusammenhang mit der Umsetzung von Nachhaltigkeitszielen diskutiert worden. Die vorhandene Umweltverträglichkeitsprüfung wäre hierbei um eine spezifische Nachhaltigkeits- und Langzeitkomponente zu erweitern (Kahl 2016b, S. 31). Eine Entscheidung, die erhebliche Auswirkungen auf zukünftige Generationen birgt, könnte im Rahmen einer solchen Prüfung systematisch hinsichtlich der Wahrscheinlichkeit des Eintretens und der Vereinbarkeit mit irreversiblen Zukunftsbelastungen untersucht werden (Kloepfer 1993, S. 37).

Eine andere Option stellt die Ernennung eines Generationenbeauftragten dar. Dieser übernimmt die Funktion eines Treuhänders, der die Interessen der zukünftigen Generationen vertritt. Hürden eines solchen Instruments liegen jedoch in seinem Spannungsverhältnis zum Demokratieprinzip (Boelling 2003, S. 446). Die zeitliche Begrenzung von Wahlperioden führt zu einer vorwiegend gegenwartsbezogenen Handlungsorientierung des Legitimierten. Klaus Ferdinand Gärditz schließt zwar eine demokratische Repräsentation zukünftiger Generationen aus, weil eine „freie Willensbildung der Repräsentierten“ nicht über eine prozedurale Legitimation rückführbar wäre (Gärditz 2016, S. 263). Deshalb sollten Zukunftsfragen demokratisch verhandelt werden.

In Betracht kommt, so unsere Einschätzung, ein speziell vom Staat bestelltes Zukunftsgremium. Ein Zukunftsgremium ist eine dauerhafte Kommission, die als Vertreter der Interessen zukünftiger Generationen in die Entscheidungsprozesse einzubinden wäre (Kloepfer 1993, S. 40). Das Gremium soll zu relevanten Entscheidungen mit langfristigen Folgen wissenschaftlich beraten.

Das Nationale Begleitgremium gemäß § 8 Standortauswahlgesetz könnte beispielsweise diese Interessenvertretung im Rahmen der Standortauswahl, Errichtung und Betrieb eines Endlagers (mit) übernehmen. Das Begleitgremium ist pluralistisch zusammengesetzt. Dessen zentrale Aufgaben bestehen in der vermittelnden und unabhängigen Begleitung des Standortauswahlverfahrens, dem Dialog und Austausch mit der Öffentlichkeit, in Empfehlungen und Stellungnahmen sowie der Vermittlung im Konfliktfall. Somit ergänzt das Gremium als dauerhaft eingerichtete Institution die Akteure im Verfahren der Endlagerung (Smeddinck 2016, § 8 Rz. 1).

Trotz fehlender subjektiver Rechte der zukünftigen Generationen müssen ihre Interessen und Belange im Rahmen der grundrechtlichen Schutzpflichten berücksichtigt werden. Aufgrund der zukunftsbelastenden Entscheidungen, die im Rahmen der Entsorgung von hochradioaktiven Reststoffen getroffen werden, besteht also Handlungsbedarf. Die Weiterentwicklung des Verfahrensrechts ist im Hinblick auf die Einbeziehung von künftigen Generationen voranzutreiben. Vorzugswürdig ist ein staatlich einberufenes Zukunftsgremium.

4 Long-term Governance unter sich auflösender Staatlichkeit

Institutionen, die für Daueraufgaben zuständig sind, brechen gelegentlich zusammen, verschwinden mitunter scheinbar, erstehen aber sehr bald wieder – eher identisch oder stark verändert –, weil sie gebraucht werden (Kromp/Lahodynsky 2006, S. 72 f.). Will man sich dem planvoller stellen, sind die Kontinuität von Fachgesetzen und Fachbehörden hervorzuheben. Der Wechsel zu einer nachhaltigeren Organisationsform einer zuständigen Organisation könnte aber auch mit Absicht herbeigeführt werden.[7]

Gesetze, die auf fachlich angemessenen Konzepten beruhen, gelten unter verschiedenen staatlichen Ordnungen fort (vgl. Oberkrome 2004, S. 520; Hönes 2015, S. 664). Ein plastisches Beispiel ist das Bürgerliche Gesetzbuch von 1900, das im Kaiserreich, in der Weimarer Republik, im Dritten Reich und in der Bundesrepublik Deutschland die zentrale Kodifikation des Zivilrechts darstellte und darstellt. Ein anderes Beispiel ist das Reichsnaturschutzgesetz von 1935, das erst 1976 auf Bundesebene durch das Bundesnaturschutzgesetz abgelöst wurde (Hönes 2015, S. 661 ff.). Manche Institutionen weisen eine noch erstaunlichere Kontinuität auf. Das dürfte im Wesentlichen daran liegen, dass

7 Vgl. bezogen auf einen anderen Kontext Reimer (2016, S. 44 f.).

die Aufgaben, derer sie sich annehmen sollen, ungelöst und dauerhaft sind (Kromp/Lahodynsky 2006, S. 72). Fachbehörden bzw. stabile handlungsfähige Arbeitszusammenhänge werden entsprechend in zukünftigen, vom heutigen Modell sich unterscheidenden Gesellschaftsordnungen benötigt. Auf dem Territorium der Niederlande existiert die erste „Wasserschaft" – eine Deicherhaltungsorganisation vergleichbar den hiesigen, ebenso traditionsreichen Deichverbänden – seit 1255 (Smolka 2021). In Deutschland bestehen Bergbehörden in Clausthal und Zellerfeld – mit Aktenbestand im Archiv aus dem Zeitraum – seit 1524 und haben gezeigt, dass ihre Arbeit notwendig und mit verschiedenen Gesellschaftssystemen kompatibel ist. Allerdings wäre genauer abzuwägen, welche institutionelle Form für die besonderen soziotechnischen Aufgaben der Entsorgung über lange Zeiträume angemessen wäre.

Die planvolle institutionelle Begleitung und „Verwaltung" eines Endlagers legt die rechtzeitige Transformation in eine Stiftung nahe. Rechtzeitig bedeutet hier eine angemessene Aufgabenübertragung oder den Wandel der Rechtsform, bevor eine staatliche Ordnung in die Krise gerät. Die Stiftung erscheint nach den bisherigen historischen Erfahrungen als eine besonders dauerhafte und über die Zeiten handlungsfähige Organisationsform: „Keine andere Institution, von privater Hand errichtet, übersteht die Jahrhunderte so unbeschadet wie die Stiftung" (Martin 2005, S. 2). So existieren in Deutschland Stiftungen, die um die 1.000 Jahre aktiv sind: Die Stiftung Vereinigte Pfründnerhäuser Münster wurde im Jahr 900 errichtet. Die Stiftung St. Johannis-Jungfrauenkloster in Lübeck stammt aus dem Jahre 1173. Seit ihrer Errichtung im Jahr 1200 arbeitet die Unterhospitalstiftung in Memmingen (Martin 2005, S. 2). Stiftungszweck ist typischerweise die Unterhaltung von Einrichtungen, die der Allgemeinheit zugutekommen.

Stiftungen sind vom Grundsatz her „unsterblich". In diesem Zusammenhang könnte das Niedersächsische Stiftungsgesetz (NStiftG)[8] angeführt werden. Nach § 6 Absatz 1 Satz 1 dieses Gesetzes ist das Stiftungsvermögen in seinem Bestand ungeschmälert zu erhalten. Die Erträge des Stiftungsvermögens sind ausschließlich für den Stiftungszweck zu verwenden (Absatz 2 Satz 1). Die Mitglieder der Stiftungsorgane sind zur ordnungsmäßigen Verwaltung der Stiftung verpflichtet (Absatz 3 Satz 1). Die Verwaltungskosten sind auf ein Mindestmaß zu beschränken (Absatz 4 Satz 1). Die Verwaltung dient der dauernden und nachhaltigen Erfüllung des Stiftungszwecks.[9] Ob und inwieweit

---

8 Vom 24. Juli 1968 (Nds. GVBl. S. 119), zuletzt geändert durch Gesetz vom 25. Juni 2014 (Nds. GVBl. S. 168).

9 Vgl. § 4 Abs. 1 Stiftungsgesetz für das Land Nordrhein-Westfalen (StiftG NRW) vom 15. Februar 2005 (GV. NRW. S. 52), zuletzt geändert durch Gesetz vom 9. Februar 2010 (GV. NRW. S. 112).

diese Vorgaben unter den Bedingungen sich auflösender Staatlichkeit eingelöst werden können, muss hier offenbleiben. Auch die genannten historischen Beispiele dürften nicht repräsentativ sein. Grundsätzlich jedoch ist dieser Organisationstypus ebenso wirksam[10] wie resilient.[11] Ein Beispiel für eine Stiftung aus Niedersachsen mit Bezug zur dauerhaften Verwahrung radioaktiver Reststoffe – wenn auch mit kürzerer Laufzeit – ist das Gesetz über die „Stiftung Zukunftsfonds Asse“ (AsseStG)[12], das darauf zielt, Belastungen in bestimmten Regionen zu kompensieren (eingehend in Weisensee 2018).

5 Weitere Ansatzpunkte und Forschungsperspektiven für Governance in sich auflösenden Zivilisationen

Neben den beiden akzentuierten Aspekten für die Ausgestaltung einer Long-term Governance sind andere Ansatzpunkte schon aufgegriffen worden oder sollten zukünftig eingehender untersucht werden:

- Das Nuclear Energy Agency (NEA) Radioactive Waste Management Committee der Organisation für wirtschaftliche Zusammenarbeit und Entwicklung (OECD) arbeitet aktuell an Möglichkeiten, über lange Zeiträume generationenübergreifend Wissen zu vermitteln und die Erinnerung wachzuhalten. Zentral ist dafür das sog. Key Information File (KIF), das eine international verwendbare, standardisierbare Struktur zur Zusammenführung der wichtigsten Informationen über jedes nationale Endlager zur Verfügung stellen soll (NEA 2015).
- Wie müsste eine resiliente Long-term Governance beschaffen sein? Dabei wäre zu untersuchen, inwieweit die von *Michael Kloepfer* genannten Kriterien für Institutionen der Langzeitverantwortung wie Stabilität und Konstanz (Langzeitpräsenz), keine oder untergeordnete Interessen in der Jetztzeit (Priorität des Langzeitinteresses), Objektivität und Akzeptanz tauglich sind und genutzt werden sollten (Kloepfer 1993, S. 36).
- Das Recht sollte so lange wie möglich in einer lebendigen Rechtskultur praktiziert werden, die nicht nur aus Personen und Institutionen besteht, sondern auch Kulturtechniken umfasst. Es kommt also auf „das je spezifische, […] historische und lokal unterschiedliche Formenarsenal, das von

---

10 In extremen Fällen bis hin zur Veränderung des Verhältnisses von Macht und Herrschaft (vgl. Schöller-Schwedes 2009, S. 171 ff.).

11 Vgl. erste grundsätzliche Erwägungen zur Resilienz der Verwaltung bei Lewinski (2016, S. 239 ff.).

12 Vom 12. November 2015 (GVBl. S. 314).

den Handelnden als Recht verstanden und praktiziert wird", an (vgl. Baer 2017, § 3 Rz. 81) und wie es dauerhaft gesichert werden kann.

- Wie lassen sich Nischen schaffen, in denen Wissen auch in schwierigen Zeiten von Katastrophen oder Beinahe-Katastrophen überdauern kann (Mulsow 2012, S. 20 f.)?
- Inwieweit könnte eine spezifische Familientradition genutzt werden, um eine archaische Form der Long-term Governance zu nutzen? Beispielsweise bewahren zwei muslimische Familien den Schlüssel des Heiligen Grabes in Jerusalem durch die Jahrhunderte. Die Grabeskirche wird als fast 1000 Jahre alte Mönchs-WG beschrieben, die von unterschiedlichen, rivalisierenden Religionsgruppen erhalten wird (Egger 2016, S. 8).
- Bietet das orale Recht unter den Bedingungen nach aufgelöster Staatlichkeit trotz seiner Fragilität Chancen, dass das Wissen über ein Endlager und seine Gefährlichkeit weitergetragen werden kann, weil es innerhalb lokaler Beziehungs- und Kommunikationsnetzwerke an einen „anthropologischen Ort", wo Menschen leben und gestalten, gebunden ist (vgl. Vesting 2011, S. 5 f.; Ehmke 2021)?

6 Fazit und Ausblick

Der Beitrag zeigt die Ausrichtung der bestehenden Rechtsordnung auf die Langfristperspektive, wie sie über einschlägige, differenzierte und bedeutsame Regelungen erfolgen kann. Im Anschluss werden der Zukunftsrat und die Stiftung als vorzugswürdige Rechtsform der Long-term Governance unter den Rahmenbedingungen sich auflösender oder wechselnder Staatlichkeiten akzentuiert. Über größere Zeiträume hinweg kann mit Blick auf die Geschichte eben nicht einfach davon ausgegangen werden, dass die staatliche Ordnung, wie wir sie kennen, langfristig stabil bleibt. Die Bundesrepublik Deutschland ist mit ihren über 70 Jahren schon die langlebigste zentralstaatliche Organisation in Deutschland seit 1871. Entsprechend wird mit Rücksicht auf die zeitliche Perspektive nach Governance-Elementen gefragt, die ihre Funktionsfähigkeit auch nach nicht nur evolutionären, sondern gar disruptiven Veränderungen der traditionellen staatlichen Ordnung behalten können. Diese Beispiele sind nur gegriffen aus einer Vielzahl anderer Ansatzpunkte, die auf Tauglichkeit und ihren Nutzen im Sinne der Long-term Governance näher untersucht werden sollten (siehe Abschnitt 5 oben).

Literatur

Baer, S. (2017): Rechtssoziologie – Eine Einführung in die interdisziplinäre Rechtsforschung. 3. Auflage, Baden-Baden

Bethge, H. (2016): Kommentierung zu § 90. In: Maunz, T.; Schmidt-Bleibtreu, B.; Klein, F.; Bethge, H. (Hg.): Bundesverfassungsgerichtsgesetz. Stand: Juli 2016, München

Birnbacher, D.; Schicha, C. (2001): Vorsorge statt Nachhaltigkeit – ethische Grundlagen der Zukunftsverantwortung. In: Birnbacher, D.; Brudermüller, G. (Hg.): Zukunftsverantwortung und Generationensolidarität. Würzburg, S. 17–33

Blumenthal, J. von (2005): Governance – eine kritische Zwischenbilanz. Zeitschrift für Politikwissenschaft 15(4), S. 1149–1180

Boelling, A.C. (2003): Ist die ökologische Generationengerechtigkeit in guter Verfassung? Vorschläge zur grundgesetzlichen Stärkung ökologischer Generationengerechtigkeit. In: Stiftung für die Rechte zukünftiger Generationen (Hg.): Handbuch der Generationengerechtigkeit. München, S. 441–470

Breuer, S. (1992): Die Gesellschaft des Verschwindens – Von der Selbstzerstörung der technischen Zivilisation. Hamburg

Brunnengräber, A.; Smeddinck, U. (2016): Möglichkeiten und Grenzen der Vereinheitlichung wissenschaftlicher Begriffe in der interdisziplinären Zusammenarbeit – eine politik- und rechtswissenschaftliche Auseinandersetzung. In: Smeddinck, U.; Kuppler, S.; Chaudry, S. (Hg.): Inter- und Transdisziplinarität bei der Entsorgung radioaktiver Reststoffe. Wiesbaden, S. 67–76

Creifelds, C. (2014): Rechtswörterbuch. München

Diamond, J. (2005): Kollaps – Warum Gesellschaften überleben oder untergehen. Frankfurt am Main

Dietrich, B.; Au, C.; Dreher, J. (2003): Umweltrecht der europäischen Gemeinschaften – Institutionen, Entwicklung und Ziele – Ein Lehr- und Arbeitsbuch. Berlin

Durner, W. (2016): Nachhaltigkeit durch Konzentration und Integration von Verfahren. In: Kahl, W. (Hg.): Nachhaltigkeit durch Organisation und Verfahren. Tübingen, S. 317–334

Egger, P. (2016): Steinheilig. Der Tagesspiegel v. 11.9.2016, S. 8

Ehmke, W. (2021): Der falsche Bodenschatz. Der Tagesspiegel v. 18.7.2021, S. 4

Ekardt, F. (2010): Das Prinzip Nachhaltigkeit. München

Friesenhahn, E. (1980): Zur Legitimation und zum Scheitern der Weimarer Reichsverfassung. In: Erdmann, K.D.; Schulze, H. (Hg.): Weimar – Selbstpreisgabe einer Demokratie: Eine Bilanz heute. Düsseldorf, S. 81–108

Gärditz, K.F. (2016): Temporale Legitimationsasymmetrien. In: Hill, H.; Schliesky, U. (Hg.): Management von Unsicherheit und Nichtwissen. Baden-Baden, S. 253–284

Hocke, P. (2013): Endlagerung hochradioaktiver Abfälle. In: Grunwald, A. (Hg.): Handbuch Technikethik. Stuttgart, S. 263–268

Höffe, O. (2015): Gerechtigkeit. München

Hönes, E.-R, (2015): 80 Jahre Reichsnaturschutzgesetz. Natur und Recht (NuR) 37, S. 661–669

Jonas, H. (1979): Das Prinzip Verantwortung. Frankfurt am Main

Kahl, W. (2016a): Der Grundrechtsschutz zukünftig Lebender. Europäisches Umwelt- und Planungsrecht (EurUP) 14(4), S. 300–312

Kahl, W.: (2016b): Einleitung – Nachhaltigkeit durch Organisation und Verfahren. In: ders. (Hg.): Nachhaltigkeit durch Organisation und Verfahren. Tübingen, S. 1–42

Kallerhoff, D. (2013): Kommentierung zu § 25. In: Stelkens, P.; Bonk, H.J.; Sachs, M. (Hg.): Verwaltungsverfahrensgesetz. 8. Auflage, München

Kloepfer, M. (1993): Langzeitverantwortung im Umweltstaat. In: Gethmann, C.F.; Nutzinger, H.G.; Kloepfer, M. (Hg.): Langzeitverantwortung im Umweltstaat. Bonn, S. 22–41

Kluth, W. (2014): Entwicklung und Perspektiven der Gesetzgebungswissenschaft. In: Kluth, W.; Krings, G. (Hg.): Gesetzgebung – Rechtsetzung durch Parlamente und Verwaltungen sowie ihre gerichtliche Kontrolle. Heidelberg, § 1

Kluth, W. (2016): Kommentierung zu Art. 20a. In: Friauf, K.H.; Höfling, W. (Hg.): Berliner Kommentar zum Grundgesetz, 51. EL, Berlin

Kromp, W.; Lahodynsky, R. (2006): Die Suche nach dem Endlager – „Make Things Small". In: Hocke, P.; Grunwald, A. (Hg.): Wohin mit dem radioaktiven Abfall? Perspektiven für eine sozialwissenschaftliche Endlagerforschung. Berlin, S. 63–82

Leggewie, C.; Welzer, H. (2010): Das Ende der Welt, wie wir sie kannten – Klima, Zukunft und die Chancen der Demokratie. 4. Auflage, Frankfurt am Main

Lersow, M.; Gellermann, R. (2015): Langzeitstabile, langzeitsichere Verwahrung von Rückständen und radioaktiven Abfällen – Beitrag zur Diskussion um Lagerung (Endlagerung). geotechnik 38(3), S. 173 – 192

Lewinski, K. von (2016): Resilienz der Verwaltung in Unsicherheits- und Risikosituationen. In: Hill, H.; Schliesky, U. (Hg.): Management von Unsicherheit und Nichtwissen. Baden-Baden, S. 239–252

Martin, J. (2005): Die Gründung einer gemeinnützigen Stiftung, Hamburg

Mulsow, M. (2012): Prekäres Wissen – Eine andere Ideengeschichte der Frühen Neuzeit. Berlin

NEA – Nuclear Energy Agency (2015): NEA Monthly News Bulletin – February 2015. https://www.oecd-nea.org/general/mnb/2015/february.html [Zugriff am 06.10.2017]

Oberkrome, W. (2004): Deutsche Heimat – Nationale Konzeption und regionale Praxis von Naturschutz, Landschaftsgestaltung und Kulturpolitik in Westfalen-Lippe und Thüringen (1900–1960). Paderborn/München

Pfordten, D. von der (2014): Normativer Individualismus. Information Philosophie 3, S. 5–15

Reimer, F. (2016): Nachhaltigkeit und „Good Governance". In: Kahl, W. (Hg.): Nachhaltigkeit durch Organisation und Verfahren. Tübingen, S. 43–61

Roßegger, U. (2011): Die Entsorgung atomarer Abfälle in der Europäischen Union – Auswirkungen der EU-Richtlinie über die Entsorgung radioaktiver Abfälle auf die Suche von Endlagern für Atommüll. Abfallrecht (AbfallR) 6, S. 276–283

Sachs, M. (2011): Kommentierung zu Art. 79. In: Sachs, M. (Hg.): Grundgesetz-Kommentar. 6. Auflage, München

Schöller-Schwedes, O. (2009): Governance durch Stiftungen. In: Botzem, S.; Hofmann, J.; Quack, S.; Schuppert, G.F.; Straßheim, H. (Hg.): Governance als Prozess -Koordinationsformen im Wandel. Baden-Baden, S. 171–200

Schreurs, M. (2017): Was macht das Nationale Begleitgremium? Transparenz und Partizipation sollen Suche nach Endlager für hoch radioaktiven Atommüll bestimmen. Umwelt aktuell 8, S. 6–7

Schuppert, G.F. (2007): Was ist und wozu Governance? Die Verwaltung 40, S. 463–511

Smeddinck, U. (2013): Rechtliche Methodik – Die Auslegungsregeln. RATUBS- Rechtswissenschaftliche Arbeitspapiere der Technischen Universität Braunschweig 4/2013

Smeddinck, U. (2016): Kommentierung zu §§ 1 und 8. In: ders. (Hg.): Standortauswahlgesetz – Kommentar. Berlin, § 1

Smeddinck, U. (2019): Feigenblatt oder Wachhund mit Konfliktradar? – Das Nationale Begleitgremium nach § 8 Standortauswahlgesetz. In: Schlacke, S.; Beaucamp, G.; Schubert, M. (Hg.): Infrastrukturrecht – Festschrift für Wilfried Erbguth. Berlin, S. 501–519

Smeddinck, U. (2021): Reversibilität in Entscheidungsprozessen – warum brauchen wir ein lernendes Verfahren? In: Brohmann, B.; Brunnengräber, A.; Hocke-Bergler, P.; Isidoro Losada, A.M. (Hg.): Robuste Langzeit-Governance bei der Endlagersuche – Soziotechnische Herausforderungen im Umgang mit hochradioaktiven Abfällen. Hamburg, S. 349–360

Smolka, K.M. (2021): Flaches Land, hohe Deiche – In den Niederlanden helfen historisch gewachsene Institutionen beim Hochwasserschutz. Frankfurter Allgemeine Zeitung (FAZ) v. 22.7.2021, S. 4

Spaemann, R. (2003): Ethische Aspekte der Endlagerung. In: Baltes, B. (Hg.): Ethische Aspekte der Endlagerung. Bonn, S. 25–36

Strack, A. (2015): Intergenerationelle Gerechtigkeit. Baden-Baden

Streffer, C.; Gethmann, C.F.; Kamp, G.; Kröger, W.; Rehbinder, E.; Renn, O.; Röhlig, K.-J. (2011): Radioactive Waste – Technical and Normative Aspects of its Disposal. Berlin

Tremmel, J. (2012): Eine Theorie der Generationengerechtigkeit. Münster

Vesting, T. (2011): Das kollektive Gedächtnis und seine Medien – Der ambivalente Status in der „westlichen Rechtstradition“. NCCR Mediality Newsletter 5, S. 3–9

Volkmer, M. (2005): Radioaktivität und Strahlenschutz. Informationskreis Kernenergie. 2. Auflage, Bonn

Voßkuhle, A. (2012): Neue Verwaltungsrechtswissenschaft. In: Hoffmann-Riem, W.; Schmidt-Aßmann, E.; Voßkuhle, A. (Hg.): Grundlagen des Verwaltungsrechts I, 2. Auflage, München, § 1

Weisensee, C. (2018): Ein Belastungsausgleich für das „Atomdreieck“ – das Gesetz über die „Stiftung Zukunftsfonds Asse“. In: Ott, K.; Smeddinck, U. (Hg.): Umwelt, Gerechtigkeit, Freiwilligkeit – insbesondere bei der Realisierung eines Endlagers für Atommüll

Willke, H. (2016): Dystopia – Studien zur Krise des Wissens in der modernen Gesellschaft. 2. Auflage, Berlin

Stefan Böschen

"Indicator politics": Non-knowledge in the context of socio-technological solutions

1 Introduction

For several decades, the problem of non-knowledge has gained more and more attention. There are two main dynamics in this. On the one hand, there is a large and multifaceted debate about science and the ignorance of science itself (cf. Ravetz 1990). In the course of this debate, not only have the principal limits of science been problematized, but also different forms of scientific ignorance have been differentiated and frameworks for their analysis have been developed (cf. Böschen et al. 2010). On the other hand, the politics of ignorance or non-knowledge became increasingly influential and visible, as the emerging debate about the precautionary principle shows. By 2000, the precautionary principle had become a principal framework for the EU's regulatory efforts (cf. EC 2000). Against this background, a policy of precaution was established in various regulatory fields. The post-release monitoring of genetically modified organisms (GMOs) in Directive 2001/18/EC or the REACH legislation are cases in point.

Nevertheless, some systematic problems remain related to such a politics of non-knowledge. In particular, the intertwined relationship between knowledge and politics, the framing of non-knowledge as a form of politics, attracts attention. This happened through the so-called agnotology studies, which pointed out that ignorance can be used strategically to avoid regulation (cf. Proctor/Schiebinger 2008). Climate change deniers are a case in point, but so are the strategies of the tobacco industry. The tobacco industry's strategy has not been to deny the relevance of studies on the relation between smoke and cancer but to claim that such relations have to been proven before any regulation. The point of this strategy is that such a level of evidence is difficult to produce for complex problems. Ultimately, this is a strategy to avoid regulation. Another line of research has been opened up by historical examples of non-knowledge (e.g., EEA 2001). With regard to different cases of "early warnings," the question was raised of how to turn past experience into future policy within the framework of the precautionary principle.

Against this background, this article aims to analyze such complex dynamics for situations where ambitious socio-technological solutions are created.

These solutions can be described as "posing a linked series of sociotechnical problems," as Paul Edwards (2004, p. 209) put it to describe their co-evolutionary development driven by physical, technical, organizational, legal, and societal factors. Therefore, the processes of articulating and solving problems need to be taken into account. In specific cases such as nuclear waste disposal, sociotechnical problems become socio-epistemic problems. This is why the parameters relevant to the technical processes have to be made objects of research. For example, in the specific context of nuclear waste disposal in geological repositories, the geological preconditions relevant to the siting and construction of a facility need to be understood. These problems are *socio*-epistemic, as the search for specific boundary conditions is highly dependent on selection processes that are influenced by path dependencies and normative preferences of basic repository concepts (cf. Smeddinck et al. 2016; Kuppler/Hocke 2019). This view of the socio-epistemic problem description and solution is further underpinned by the fact that nuclear waste disposal is in principle a process over time spans that are unprecedented for governance procedures. Decisions made in the present reach far into the future – a high-level nuclear waste repository should remain safe and secure for up to 1 million years. Therefore, at least partial reversibility of problem descriptions and problem-solving measures would be beneficial for safety reasons as well as for future societies to decide on their preferred way of dealing with the waste. However, such reversibility depends both on the clarity of the formulation of socio-epistemic/-technical problems and on the transparency and legitimacy of the institutions involved to deal with conflicts between the different knowledges articulated to describe and solve the problem of nuclear waste disposal.

The starting assumption of the article is that these processes can be described as the formation of knowledge regimes in which the respective problems are shaped in their form and structure while evidence is built to solve them. Since non-knowledge and its construction is a major driving force in such processes, the form of arguing with non-knowledge and its consequences for the formation of problems is presented here (Sect. 1). In a second step, the social side of these socio-epistemic dynamics is examined, arguing for a specific concept of knowledge regimes for understanding the dynamics of ambitious socio-technological solutions. The argument is that we need to improve our analytical tool box for understanding knowledges by using an approach that reflects the use of indicators in problem solving processes. And it is proposed to differentiate between criteria, indicators, and observables (Sect. 2). In relation to this analysis, I will proceed with the argument to analyze the processes of (non-)knowledge formation in knowledge regimes as indicator politics. Indicator politics is a form of knowledge politics that presents specific aspects of the problem description and solution, focusing attention on selected indicators.

Any politics that is highly knowledge-based and knowledge-dependent is necessarily indicator politics. But, the key question is whether there are legitimate and effective institutionalized procedures for managing the conflicts between different knowledges by transparently sorting indicators and their observables (Sect. 3). I will end with some conclusions (Sect. 4).

2 Non-knowledge: not only the other side of scientific knowledge

Over the last 30 years, a lively debate about forms, perspectives, and consequences of non-knowledge or ignorance has developed (cf. for many: Gross/McGoey 2015). This multifaceted debate was mainly inspired by the central insight, coined as early as in the 1980s, that non-knowledge is a social construct embedded in and shaping social interactions (Smithson 1985). This argument is also the starting point here, but in a specific way and intention. By whom is non-knowledge defined and to what end? In relation to science, non-knowledge was seen as specified ignorance (cf. Merton 1987). Here, the frame of reference is a scientific discipline defining non-knowledge according to rules established by themselves. This situation is quite easy to handle, since the relevant practices and perspectives belong to a defined group of actors. But, as we know from attempts at interdisciplinarity, the situation becomes more difficult when different disciplines try to define non-knowledge together. Nevertheless, the actors agree on general scientific standards. This offers the possibility of division of labor and thus of working out the problem.

However, when it comes to typical socio-epistemic problems such as climate change, the energy transition, nuclear waste disposal, or other grand challenges, the definition and solution of such problems is part of a multifaceted process. Since heterogeneous values and interests are at stake and the knowledge resources are typically contested, such processes have to be carried out in a transdisciplinary manner. This seems to be important because, due to this situation, parts of society demand to be involved in defining what is part of the problem and need to change their behavior as part of the solution. With regard to such transdisciplinary endeavors, the claim of non-knowledge becomes most difficult. For this reason, not only are disciplinary standards for framing non-knowledge conflicting, but the general epistemic standards are also up for debate. Therefore, the general way of defining and solving problems is part of the debate as well. And here, a broader set of relevant strategies is up for discussion. Then there is not only the well-known expert dilemma of different experts struggling for dominance in the debate, with the consequence of delegitimizing expertise in itself. The problem is more subtle and relates to the question of how epistemic standards of knowledge in defining problems can be made evident. Therefore, the thesis here is that the debate about non-knowledge is primarily one fueled by the heterogeneity of knowledge resources and the resulting challenges of constructing problems and building evidence to solve them.

This can be exemplified by looking at how different perspectives and disciplines are relevant to the construction of different epistemic challenges related to nuclear waste disposal. More precisely, there are two indications for this view. *Firstly*, an important aspect in this field of problem construction are specific conceptual frameworks that provide an overarching strategy for sorting and evaluating the different knowledge sources in the field. These include, for example, the so-called FEP catalogs (Schneider et al. 2017, p. 70): *F* stands for *features* that characterize the state of a nuclear waste disposal site at a given time. *E* stands for *events* that can influence the state of a disposal site in a short time span (e.g., earthquakes). *P* stands for *processes* that can affect the state of the disposal site over a longer time span (e.g., climate change). There are different ways of defining such FEP catalogs, but in any case they sort out the heterogeneity of knowledge relevant to nuclear waste disposal. *Secondly*, there are different knowledges. For example, there are different ways of modeling transport and exposure – and they depend on their disciplinary background. Overall, it is known in the field that there are different knowledges and the need to collect, sort, and structure these sources in order to define the problem by building evidence.

To go a step further, I would like to refer to the scheme of the so-called *argumentum ad ignorantiam*. Figure 1 shows the main structure of the argument. If we look at the form of the argument, we can see that this conclusion is quite vulnerable to attack. This is related to three aspects. *Firstly*, there is the correlation between "being dangerous" and "showing aligning risky effects." This alignment depends on correlations between two states. Such correlations are nothing but simple. On the contrary, this field is quite open and is therefore a space for wild speculation and strong causal relationships. *Secondly*, there is the statement about observation. This is highly dependent on the established routines of observation. For example, with regard to the case of CFCs (chlorofluorocarbons), accused in the 1970s of destroying the stratospheric ozone layer, this substance was extensively tested in the 1930s using all available testing routines. But as they focused on short-term effects, they did not help foresee the destruction of the ozone layer, which is a long-term and distant process. Thus, the statement of observation can be attacked systematically by offering new observation possibilities. *Thirdly*, the conclusion of "not being dangerous" or, more far-reaching, "safe" is a specific qualification of the state of affairs with regard to social norms. Nevertheless, norms can change and with them also the qualification of what is to be considered dangerous/safe – and what is not. As mentioned before, the conclusion is thus based on a number of assumptions, and there are different starting points for controversy.

Figure 1: Argumentum ad ignorantiam

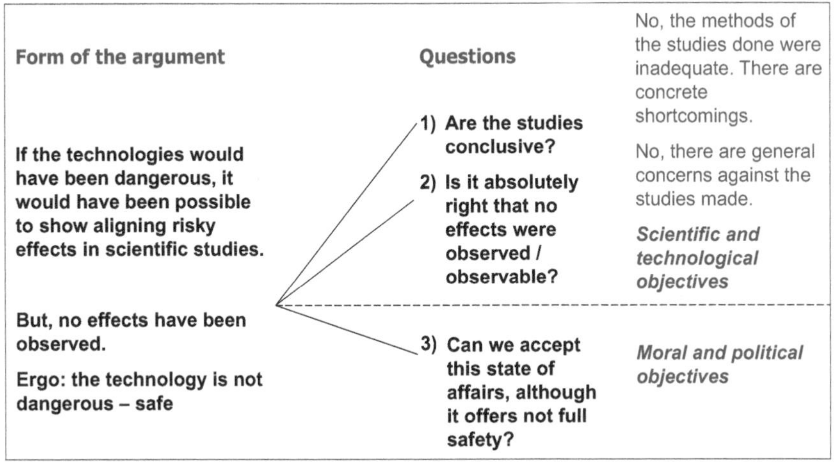

Source: Adapted from Soentgen 2015, p. 141

If we look at the second column of the chart, there are two parts. One part relates to scientific and technological objectives, the other to moral and political objectives. In practice, it might be difficult to differentiate between the two, but this model is an analytical one to make clear that there are in principle two different layers of arguments. With regard to scientific and technological objectives, one can ask, for example, whether the studies conducted can be considered conclusive. In many cases, the available studies address only parts of the overall problem. Therefore, questions about the missing parts of the picture are likely and important. The second exemplary question relates to the fact that the observations were not yet precise enough or did not yet have the right measurement technology or the relevant perspective. In the face of complex problems, these kinds of questions are quite likely, as different disciplines struggle within their own field of expertise, but also among them, over which observation strategies are appropriate and which are not.

On the other hand, there are moral and political questions. Typically, they take the form of asking whether or not a specific state of affairs can be considered conclusive for the decision about a technology. These kinds of questions are quite difficult to answer as they relate to the necessarily limited knowledge base about a problem. Therefore, these questions address both the definition and the solution strategies of what can and should be considered *the* problem. It is about shaping the problem. And shaping a problem is always done in the light of specific norms and values. Moreover, these can be contested. Viewed in this way, the epistemic difficulties in shaping the problem turn into uncertainty about the norms and values by which the decision is made. In summary, there are several issues regarding the conclusion of this argument that give rise to discussion. Moreover, the most disturbing effect may be that it is quite difficult to bring a debate to a close under such conditions. Therefore, it is possible to relate more or less any observation to the technology under debate and thus to reopen the debate again and again (cf. Bora 2002). Once opened up, the end of such a debate seems to depend more on the attention cycles of public debates than on the conclusiveness of arguments in the debate.

These questions are highly relevant in terms of analytics for understanding processes of socio-epistemic problems, their definition, and solution. The cognitive framing of such problems in many cases follows the logic of this argument. Firstly, an important part of defining a problem is to define what is *not* part of the problem. And these arguments are put forward along the lines of, "If this aspect must be part of the problem, then the problematic quality of this aspect should have been pointed out. But this was not the case. Therefore, this aspect is not relevant." Secondly, the logic of problem processing follows the path of problem expansion. Then, arguments of the above structure are put forward by the opponents to provide reasons for expanding the horizon of the problem to be defined. The consequence in the first case is a restriction of the problem definition, in the second case an expansion of the scope of the problem. These general arguments are related to the structure of the knowledge base of the problem. To understand the concrete dynamics of problem construction and solution, we need to look more closely at the social structure of this process.

3 Ambitious socio-technical solutions as regimes of knowledge

In the analysis of the social structure, I would like to argue for looking at knowledge regimes. This concept has been developed in different disciplinary arenas, especially in political science with a focus on formal (mostly legal) contracts in transnational cooperation (cf. Kohler-Koch/Schaber 1994), in sociology with an emphasis on imbalances of discourses and focus on power relations (Pestre 2003; Wehling 2007), and finally in the Science & Technology Studies in the concept of civic epistemology (cf. Jasanoff 2005). In Sheila Jasanoff's version, a civic epistemology "[...] refers to the institutionalized practices by which members of a given society test and deploy knowledge claims used as a basis for making collective choices" (Jasanoff 2005, p. 255). I would like to refer to these debates by pointing out one argument. This argument relates to the fact that in most of these definitions the question of knowledge and its specific forms is rarely addressed (for an exception see Jasanoff 2004). Therefore, in order to analyze the co-production process of knowledge and social structure as a *process*, a new conceptual vocabulary seems to be important. Against this background, I would like to put forward two arguments. Firstly, there is the question of how to conceptualize the dynamic aspect of knowledge production in knowledge regimes. This question will be addressed using the concept of socio-technical or rather socio-epistemic problems. Secondly, there is the question of whether we have an analytics of knowledge that allows us to understand the concrete structure of knowledges, their use, and development in problem-solving processes. In this context, I propose a heuristic for differentiating such knowledges.

With regard to the first question, we need a starting point for including the aspect of socio-epistemic dynamics in the concept of knowledge regimes. The idea is to combine the above definitions with a proposal by Paul Edwards in analyzing infrastructures. He described infrastructures "as posing a linked series of sociotechnical problems" (Edwards 2004, p. 209). By analogy, one can argue that such a chain of problems to be solved drives the dynamic of debate in knowledge regimes and forms its structure. But these problems are socio-epistemic in that the relevant standards of evidence must be defined and the methods of solving a problem need to be determined. This discoursive dynamic of defining problems and their related solution mechanisms must be seen as both a cognitive and an institutional process. Therefore, I suggest the definition of knowledge regimes as discoursively constructed and institutionally stabilized spaces of practices for articulating, discussing, and solving socio-epistemic problems.

The second question aims to analyze the construction of socio-epistemic problems using different knowledges. Typically, proponents and opponents base their arguments on different sources of knowledge. But how can the relevant differences between these arguments be made visible? I would like to suggest an analytical framework to classify the different aspects of knowledge used for problem description. This proposal differentiates knowledge sources by looking at three aspects: criteria, indicators, and observables (Böschen 2014, pp. 40 f.). Criteria evaluate indicators against the background of main cultural values or interests (e.g., "safe disposal," "precautionary principle," or "low-carbon society"). Indicators represent an effect-related aspect of a problem that should be considered or solved (e.g., "radiotoxicity," "cancerogeneity," "CO_2 footprint"). Finally, observables apply indicators by providing specified methods for empirical observations or testing strategies (e.g., "dosimetry" or "LC50 test" for acute toxicity).

Against this background, we can take a closer look at the dynamics and logics of articulating and solving socio-epistemic problems constituting knowledge regimes. The first and most important aspect is that of "problem setting." The dynamic within such regimes of knowledge is fueled by the opening up of problems and the processes associated with such an affordance, which results simply from putting forward a problem. This means that it is possible to occupy a knowledge regime by "problem setting." This can be studied in numerous processes. Without naming phenomena of this kind that way, they were nevertheless analyzed in the so-called agnotology studies (cf. Oreskes/Conway 2010).[1] For example, with regard to the strategies of the tobacco industry to avoid regulation, it was observed that the proponents' aim was not to argue that cigarettes had no health effects. The strategy was more subtle. They claimed that if there was a link between cigarettes and cancer, it would have to be considered a severe health hazard that needed to be regulated. Therefore, this conjecture should be thoroughly investigated. If strong evidence of smoking as a cause of cancer were found, then cigarette use should be regulated. The point here was to demand an unambiguousness that is hard to achieve. It took about 50 years of research to accomplish this. The proposed knowledge analytics can be applied to reconstruct this process. Although there was consensus on the criteria ("protect humans from health hazards") and indicators ("cancerogeneity"), the problem was shifted to the level of observables. And since there was a call for "high evidence," many data points had to be created and various observation methods were questioned. This takes time.

A second very important aspect for the study of knowledge production in knowledge regimes is that of selecting problem descriptions. Problem descriptions change over time in knowledge regimes. In any case, knowledge regimes regulate which problem descriptions are feasible and legitimate – and which are not. Therefore, other problems and aspects are necessarily put aside or ignored. Problem dynamics within knowledge regimes is a story of selecting problems in the expectation of reducing the non-knowledge about problems put into the center of the regime. This means, for example, that there are institutions that define specific temporal spaces for reducing non-knowledge. A case in point is the establishment of so-called post-release monitoring in

---

1 This new field of agnotology studies is part of the debate about scientific non-knowledge. It highlights the making and unmaking of facts and truths in debates on knowledge politics. Agnotology studies aim at understanding the cultural, social, and economic forces behind the "interested" production of ignorance. Cases in point are the tobacco industry or climate change deniers (cf. Proctor/Schiebinger 2008; Oreskes/Conway 2010; Proctor 2012).

the European Union to regulate GMOs (genetically modified organisms). This institutionalizes a 10 year period during which possible negative effects of the authorized organism can be monitored. After this time span, a new approval of the GMO by the authority is required. The administration can therefore decide on the basis of the new state of affairs whether the authorization is appropriate – or not. This strategy of temporalization is very important to solve non-knowledge problems (cf. Böschen et al. 2010). However, this strategy depends on whether non-knowledge is seen as temporally resolvable. When non-knowledge related to the main problems of a regime proves irreducible, the social balance within the regime becomes fragile. This is the case with regard to nuclear waste disposal as there are time spans relevant to regulation that are typically not addressable by scientific evidence (cf. Schmidt et al. 2017). Against this background, the struggle over specific non-knowledge and its temporal stability is highly relevant for the social order and stability of a knowledge regime.

To draw together the different strands of argumentation: Ambitious socio-technical solutions can typically be understood as knowledge regimes. There are many examples of this, such as the energy transition, the mobility transition, or nuclear waste disposal. In all these cases, changes in social and epistemic orders are highly intertwined. Key characteristics are the complexity of the infrastructures associated with these solutions, the multifaceted history of problem solving, the existence of multiple futures and development pathways (Lösch/Schneider 2016), the landscapes of multiple knowledges and non-knowledges (Böschen et al. 2010), and finally specific social challenges such as trust or confidence, which can be interpreted as "socio-technical problems" (Büscher/Sumpf 2015). An important aspect with regard to the parallelism of different knowledge regimes is their interrelatedness. Especially in the case of energy policy, we observe in Germany a highly relevant coupling of the energy transition with a transition of the mobility system toward e-mobility. Therefore, overarching questions of problem solving generally arise simultaneously. Thus, within knowledge regimes – as regimes of specifying and ignoring problem-solving processes – both the boundary conditions of problem solving and the objects of problem solving are constituted while processing problems. Against this background, I would like to transfer these conceptual arguments to the analysis of what I call indicator politics and relate them to the debate on nuclear waste disposal.

4 Not only arguments, but ... indicator politics

Indicator politics is a form of knowledge politics by presenting specific aspects of the problem description and solution while focusing attention on selected indicators. In this way, certain parts of the problem-solving process are highlighted while others are more or less ignored. A case in point is the *Ecomodernist Manifesto* (Asafu-Adjaye et al. 2015). We can see how its argumentation is based on presenting specific indicators to describe the problem and then coming to a conclusion on what action needs to be taken.

> "Transitioning to a world powered by zero-carbon energy sources will require energy technologies that are power dense and capable of scaling to many tens of terawatts to power a growing human economy. Most forms of renewable energy are, unfortunately, incapable of doing so. The scale of land use and other environmental impacts necessary to power the world on biofuels or many other renewables are such that we doubt they provide a sound pathway to a zero-carbon low-footprint future [...]."

And the consequence of setting the scene in this way is:

> "Nuclear fission today represents the only present-day zero-carbon technology with the demonstrated ability to meet most, if not all, of the energy demands of a modern economy. However, a variety of social, economic, and institutional challenges make deployment of present-day nuclear technologies at scales necessary to achieve significant climate mitigation unlikely. A new generation of nuclear technologies that are safer and cheaper will likely be necessary for nuclear energy to meet its full potential as a critical climate mitigation technology" (Asafu-Adjaye et al. 2015, p. 22 f.).

Looking at the quote from the Ecomodernist Manifesto, we can see that their narrative is built on a set of selected indicators and the criterion of a zero-carbon low-footprint future. Such narratives make it possible to make sense of the perception of the world, symbolizing social order and highlighting appropriate action. And the dynamics of knowledge regimes, as well as the alignment deficits, can be understood by looking at the "indicator politics" that unfold in the course of public debate and regulatory action. Here we see that the indicator "power density" is used to describe the quality of energy production and "land use" is used to describe the impacts of technologies; the conclusion regarding the solution is then obvious for the authors. I would argue that the use of selected indicators is by no means accidental. On the contrary, this selection is a political strategy. This is why the problem description highly depends on the use of indicators. By putting forward certain indicators, it is possible to steer the debate toward a perspective of interest – wherever this interest is routed in.

Therefore, knowledge is needed about the forms and practices of institutional procedures to get an overview of the variety of indicators, to make a transparent selection of indicators, to collect and maintain the related information, and finally to continuously adjust the indicators during the problem-solving process. It is thus no coincidence that in the debate on nuclear waste disposal, the notion of "long-term governance" is highly related to both the problems of democratic decision making and the problems of learning while doing waste disposal. Moreover, the boundary conditions for such learning are much more difficult as they change over time. Consequently, Sophie Kuppler and Peter Hocke argue for adopting and specifically reformulating the concept of "stewardship" for this type of sites. In such a reformulated form of stewardship, the steward "would be at the same time practically capable of intervening at the site and maintaining and furthering knowledge on how to intervene" (Kuppler/Hocke 2019, p. 1351). The key question here is which knowledges are relevant – i.e., which indicators are used in creating a knowledge landscape for waste disposal – and how knowledge can be produced by using specific observables – i.e., which empirical strategies are related to the selected indicators.

Since the processes of "doing waste disposal" are located in a highly contested field, the selection of indicators and observables is a key feature of any institutionalized form of problem solving. It can be said that the legitimacy and efficacy of the problem-solving process mainly depends on the accuracy and transparency of the institutionalized problem-solving procedure for selecting indicators and operationalizing them through selected observables. If this process fails, nuclear waste disposal will also fail.

5 Conclusions

We can conclude that ambitious socio-technological solutions challenge the processes of problem construction behind them. While driving the development of socio-technological solutions, specific indicators are put forward to describe their relevance and side effects. Within this process of indicator selection, the whole challenge of socio-epistemic problem solving can be analyzed. The respective indicators are put forward by actors routed in their disciplines and organizational backgrounds. In their worldview, the use of selected indicators is common sense. However, this is not necessarily true for other (collective) actors and their worldviews. The struggle over indicators and their empirical measurement therefore shows the heterogeneity of knowledge perspectives and the density of conflicts.

However, besides looking at the indicators and their observables, it is crucial to look closely at the social dynamics related to them. Who are the (collective) actors putting forward which indicators and why? The process of constructing and solving problems is not only discursive but also related to institutions and practices. Knowledge regimes and their evolution are highly dependent on this relation between discursive and institutional dynamics. For this reason we can analyze these dynamics as indicator politics. Selected indicators specify socio-epistemic problems. The scope of the indicator set is related to the scope of non-knowledge remaining open. Moreover, the struggle over the indicators and their observables is a struggle over the problem structure and responsibilities for who has to contribute which knowledge resources for solving the problems then defined.

References

Asafu-Adjaye, J.; Blomqvist, L.; Brand, S.; Brook, B.; DeFries, R. et al. (2015): The Ecomodernist Manifesto. http://www.ecomodernism.org [Accessed 21.09.2021]

Bora, A. (2002): Ökologie der Kontrolle. Technikregulierung unter der Bedingung von Nicht-Wissen. In: Engel, Chr.; Halfmann, J.; Schulte, M. (eds.): Wissen – Nichtwissen – Unsicheres Wissen. Baden-Baden, pp. 254–275

Böschen, S. (2014): Opening the Black Box: Scientific Expertise and Democratic Culture. In: Michalek, T.; Hebakova, L.; Hennen, L.; Scherz, C.; Nierling, L.; Hahn, J. (eds.): Technology Assessment and Policy Areas of Great Transitions. Prague, pp. 37–47

Böschen, S.; Kastenhofer, K.; Rust, I.; Soentgen, J.; Wehling, P. (2010): The Political Dynamics of Scientific Non-Knowledge. Science, Technology & Human Values 35(6), pp. 783–811

Büscher, C.; Sumpf, P. (2015): "Trust" and "confidence" as socio-technical problems in the transformation of energy systems. Energy, Sustainability and Society 5(34); DOI: 10.1186/s13705–015–0063–7

EC – European Commission (2000): Communication from the Commission on the precautionary principle.Com (2000) 1 final. Brussels

Edwards, P.N. (2004): Infrastructure and modernity: Force, time and social organization in the history of sociotechnical systems. In: Misa, T.; Brey, P.; Feenberg, A. (eds.): Modernity and Technology. Cambridge, MA, pp. 185–226

EEA – European Environment Agency (2001): Late lessons from early warnings: the precautionary principle 1896 – 2000. Environmental issue report, No. 22, Copenhagen

Gross, M.; McGoey, L. (eds.) (2015): Routledge International Handbook of Ignorance Studies. London

Jasanoff, S. (ed.) (2004): States of knowledge: the co-production of science and the social order. London/New York

Jasanoff, S. (2005): Designs on Nature. Princeton

Kohler-Koch, B.; Schaber, Th. (1994): Regimeanalyse. In: Kriz, J.; Nohlen, D.; Schultze, R.O. (eds.): Lexikon der Politik, Bd. 2: Politikwissenschaftliche Methoden. Munich, pp. 402–404

Kuppler, S.; Hocke, P. (2019): The role of long-term planning in nuclear waste governance. Journal of risk research 22(11), pp. 1343–1356

Lösch, A.; Schneider, C. (2016): Transforming power/ knowledge apparatuses: the smart grid in the German energy transition. Innovation 29(3), pp. 262–284; http://dx.doi.org/10.1080/13511610.2016.1154783 [Accessed 21.09.2021]

Merton, R. K. (1987): Three Fragments from A Sociologist's Notebook: Establishing the Phenomenon, Specified Ignorance, and Strategic Research Materials. Annual Review of Sociology 13, pp. 1–28

Oreskes, N.; Conway, E. (2010): Merchants of Doubt. New York

Pestre, D. (2003): Regimes of Knowledge Production in Society: Towards a more political and social reading. Minerva 41(3), pp. 245–261

Proctor, R. (2012): Golden Holocaust. Origins of the Cigarette Catastrophe and the Case for Abolition. Berkeley, CA

Proctor, R.; Schiebinger, L. (eds.) (2008): Agnotoloy. The Making und Unmaking of Ignorance. Stanford

Ravetz, J.R. (1990): The merger of knowledge with power. Essays in critical science. London

Schmidt, G.; Kallenbach-Herbert, B. with Minhans, A.; Alt, S. (2017): ENTRIA-Arbeitsbericht-09 "Endlagerung ohne Vorkehrungen zur Rückholbarkeit – Technik- und Sicherheitsaspekte." ENTRIA

Schneider, M.; Froggatt, A. with Hazemann, J.; Katsuta, T.; Ramana, M.V.; Rodriguez, J.C.; Rüdinger, A.; Stienne, A. (2017): The World Nuclear Industry. Status Report 2017. Paris

Smeddinck, U.; Kuppler, S.; Chaudry, S. (eds.) (2016): Inter- und Transdisziplinarität bei der Entsorgung radioaktiver Reststoffe: Grundlagen – Beispiele – Wissenssynthese. Wiesbaden

Smithson, M. (1985): Toward a Social Theory of Ignorance. Journal for the Theory of Social Behaviour 15(2), pp. 151–172

Soentgen, J. (2015): Argumentieren mit Nichtwissen: Die Risikodiskurse über Mobilfunk und Grüne Gentechnik. In: Wehling, P.; Böschen, S. (eds.) (2015): Nichtwissenskulturen. Baden-Baden, pp. 119–156

Wehling, P. (2007): Wissensregime. In: Schützeichel, R. (ed.): Handbuch der Wissenssoziologie und Wissensforschung. Konstanz, pp. 704–712

Oliver Sträter

Bedeutung menschlicher Faktoren für eine dauerhafte Sicherheit von Entsorgungsoptionen

1 Einleitung

Die Sicherheit von menschengemachten Systemen hängt naturgemäß von Menschen ab. Das gilt auch für Systeme zur Entsorgung radioaktiver Abfälle. Selbst eine langfristige Entsorgungsoption ist letztendlich die Konsequenz einer menschlichen Entscheidung und Bewertung, dass eine Lagerung in einer bestimmten geologischen Konstellation mit bestimmten technischen Behältnissen geeignet ist und Bestand hat. Des Weiteren ergibt sich durch die Langfristigkeit der Entsorgung radioaktiver Abfälle die Anforderung an die Betreiberorganisation und das Umfeld dieser Organisation, im Bedarfsfall auf Sicherheitsprobleme zu reagieren.

Im Kontext einer dauerhaften Sicherheit von Entsorgungsoptionen stellt sich die Frage, wie über lange Zeiträume hinweg ein solches Entsorgungssystem auf einem angemessenen Sicherheitsniveau gehalten werden kann und wie verhindert werden kann, dass Fehler oder Fehlentwicklungen nicht oder zu spät erkannt werden und ggf. Kontaminationen der Umgebung des Endlagers stattfinden. Hier spielt das menschliche Verhalten, insbesondere hinsichtlich eines langfristigen, selbstkritischen Hinterfragens, ob sich das Entsorgungssystem (also dessen Technik, Organisation und Organisationsumfeld) in einem sicheren Zustand befindet, eine entscheidende Rolle.

2 Grundlagen eines selbstkritischen Hinterfragens

2.1 Aufgaben versus zielbezogenes menschliches Verhalten

Üblicherweise und durch das Rechtssystem entsprechend unterstützt werden normative Anforderungen an ein sicheres System gestellt. Das normative System stellt Anforderungen an den Betreiber oder Bediener, die dieser zu erfüllen hat, um einen sicheren Zustand des Systems zu gewährleisten. Sicherheitskritische Zustände werden über die Nichteinhaltung des normativen Systems definiert und geahndet.

Diese Sichtweise funktioniert in einfachen Systemen hinreichend und kann Grundlage der sicheren Gestaltung sein. Je komplexer Systeme werden, je mehr Zielvorgaben an das System gemacht werden oder je mehr Wechselbeziehungen in einem System vorherrschen, desto schwieriger wird es jedoch, das normative Verhalten als ausschließlich die Sicherheit gewährleistend zu betrachten (Reason 1990).

Das bekannteste Beispiel ist das Thema Sicherheit versus Kosten, wie es von Hollnagel (2009) zusammengefasst dargestellt wird und insbesondere in der Verkehrstechnik (Flugbereich, Straße, Schiene) ein offensichtliches Problem darstellt. Ein typisches Beispiel, das jeder aus eigener Erfahrung kennt, ist folgendes: Jeder Autofahrer möchte pünktlich und sicher von A nach B reisen und macht sich – mehr oder weniger explizit – einen entsprechenden Plan für die Abfahrts- und Ankunftszeit einer Fahrt. Kommt es dann zu Verzögerungen während der Fahrt (z. B. Stau aufgrund hohen Verkehrsaufkommens), ändert sich das Zielverhalten: Die zu Beginn der Fahrt ausgewogenen Ziele „Sicherheit“ und „Pünktlichkeit“ verschieben sich zugunsten des Ziels Pünktlichkeit; Sicherheit wird als „Zielopfer“ toleriert; d. h. man fährt unsicherer, also auch schon mal bei gelb über die Ampel oder mit überhöhter Geschwindigkeit oder führt kritische Überholmanöver durch, um so den Zeitverlust auszugleichen.

Allgemein gesprochen führen Zielkonflikte psychologisch gesehen zu begrenzt rationalen Zieloptimierungen bei den dem Zielkonflikt ausgesetzten Personen (Kahnemann/Tversky 1979). Personen in Zielkonflikten ändern ihr Verhalten von einem eher normativen, aufgabenorientierten zu einem zielbezogenen Verhalten. Aufgabenbezogenes Verhalten beschreibt die konkreten menschlichen Handlungen in einem Anforderungsszenario und geht davon aus, dass der Mensch diese Handlungen nach bestem Wissen und Gewissen erfolgreich durchführen möchte. Sicherheitstechnisch wird bewertet, inwiefern der Mensch eine ihm gestellte Aufgabe wegen Ablenkung oder ungünstiger Bedingungen nicht erfüllen kann (Swain/Guttmann 1983). Zielbezogenes Verhalten beschreibt das Phänomen, dass im Umfeld der Aufgabe weitere Anforderungen erfüllt werden müssen und nicht ausschließlich ein ganz bestimmtes Anforderungsszenario (VDI 4006 Blatt 2, 2017). Von der handelnden Person sind hier Entscheidungen und Priorisierungen von Aufgaben und Handlungen erforderlich, die dazu führen, dass die geforderte Handlung unter Umständen gar nicht oder erst mit langer Verzögerung durchgeführt wird (Mosneron-Dupin et al. 1997). Der Zielkonflikt kann auch so aufgelöst werden, dass die Person eine ganz andere Handlung als die normativ vorgesehene durchführt, um einen aus ihrer Sicht optimalen Kompromiss zwischen den unterschiedlichen Zielstellungen zu schaffen.

In frühen Betrachtungen des menschlichen Einflusses auf die Sicherheit von Systemen wurden dagegen ausschließlich normative, aufgabenbezogene

Bewertungen menschlicher Handlungen durchgeführt (Überblick über die Methoden z. B. in Swain 1989). Die Erfahrung schwerer Störfälle wie Tschernobyl oder Fukushima zeigen jedoch die Bedeutung der sogenannten zielbezogenen Sichtweise (Sträter et al. 1999; Apostolakis et al. 2004), denn bei all diesen schweren Störfällen spielen durchweg zielorientierte Verhaltensweisen eine entscheidende Rolle.

Zielbezogenes Verhalten wirkt insbesondere unter Randbedingungen, bei denen komplexe Zielabgleiche erforderlich sind. Übertragen auf die dauerhafte Sicherheit von Entsorgungsoptionen sind u. a. folgende Aspekte relevant (vgl. Sträter 2005):

Zeitlicher Ablauf der Situation:

- Plötzlichkeit des Eintritts einer Störung: Dies führt dazu, dass zunächst verweigert wird, das Ereignis anzuerkennen.
  Beispiel: Ein spontanes Versagen eines Behälters im Endlager sorgt für den Entscheidungskonflikt, ob nun eine Rückholung stattfinden soll oder nicht. Hier ist zunächst mit Reaktanz zu rechnen: „Ist das wirklich erforderlich wegen eines Behälters, den Rückholprozess zu starten?“
- Schleichende Prozessentwicklung, lange ereignislose Phase vor Störungseintritt: Dies führt dazu, dass die Entwicklung als normal und nicht kritisch eingeschätzt wird.
  Beispiel: Durch langsame geotektonische Veränderungen des Endlagers verschieben sich die Geometrien des Stollens und über Jahre ist scheinbar keine Handlung erforderlich; der Betrieb „gewöhnt“ sich daran, dass die Geometrie immer ungünstiger wird, bis diese eines Tages soweit verschoben ist, dass Zuwege für die Rückholung versperrt sind.
- Positive Erfahrung mit dem System und Übervertrauen in das System.
  Beispiel: Das Monitoringsystem zur Messung der Umgebungsaktivität im Stollen ist hochzuverlässig und hat über Jahre korrekt Strahlungsfreiheit gemessen. Trotz widersprüchlicher Information (z. B. leichte Zunahme der Aktivität im Grundwasser) wird an der Korrektheit des Messsystems nicht gezweifelt.
- Verweigerung der Aufgabenwahrnehmung kurz vor Beendigung einer Aufgabe, im Luftfahrtbereich als „home base syndrome“ bekannt.
  Beispiel: Nach Abschluss der offiziellen Rückholphase und kurz vor dem endgültigen sicheren Einschluss werden Werte gemessen, die zur Rückholung auffordern, aber die Organisation hat sich innerlich schon auf den sicheren Einschluss vorbereitet und ist aus politischen oder technischen Gründen nicht mehr Willens oder in der Lage, die Rückholung vorzunehmen.

Systemzustände in der Situation:

- Zusätzlich anstehende Aufgaben im System.
  Beispiel: Eine nach vorher festgelegten Kriterien erforderliche Rückholung wird trotz Bedarf aufgrund von Wirtschaftlichkeitsbetrachtungen nicht vorgenommen.
- Multiple Fehler oder Probleme wie dominante Zusatzstörung, Teilausfall des Systems.
  Beispiel: Politische oder gesellschaftliche Krisen überschatten die technische Anforderung der Rückholung und führen zu Vermeidungsstrategien und Verschiebungen, die dann so lange anhalten, bis es zur Freisetzung von Strahlung kommt.
- Mangelnde Transparenz von Abhängigkeiten im System.
  Beispiel: Das Lagerungskonzept legt geringe Scherkräfte an den Behältern zugrunde und die Behälter werden nach diesem Konzept ausgelegt; in der tatsächlichen geotektonischen Bewegung treten dann hohe Scherkräfte auf, die den Behälter beschädigen.
- Ambivalente Symptome, die mehrere Schlussfolgerungen zulassen.
  Beispiel: Die Messsyteme liefern widersprüchliche Informationen und dem Messwert, der den geringsten technischen Aufwand bedeutet, wird am meisten Glauben geschenkt.
- Vermeidung relevanter Arbeiten wegen drastischer Aufwände.
  Beispiel: Eine erforderliche Rückholung im Schadensfall wird aufgrund des technischen, organisatorischen und/oder politischen Aufwands nicht vorgenommen.

Liegen solche Randbedingungen vor, neigt der Mensch dazu, die klaren aufgabenbezogenen Anforderungen nicht oder nicht hinreichend anzugehen; zur Lösung des Problems werden Vermeidungs- und Optimierungsstrategien entwickelt. Diese Vermeidungs- und Optimierungsstrategien werden in der Kognitionspsychologie als Heurismen bezeichnet (Kahnemann/Tversky 1979).

Sie führen dazu, dass der Mensch oder die Organisation bestimmte sicherheitsgerichtete Aufgaben nicht mehr hinterfragt und sich für eine vermeintlich funktionierende Lösung entscheidet, auch wenn diese risikobehafteter ist oder sogar in ein Schadensereignis hineinläuft (OECD 2004). Bezogen auf die Systemsicherheit haben solche Lösungen die negative Eigenschaft, selbst ausgeklügelte Sicherheitsmechanismen, Barrieren oder administrative Vorkehrungen außer Kraft zu setzen: Ist der handelnde Mensch überzeugt, dass das von ihm verfolgte Ziel das „richtige" ist oder der von ihm entwickelte Kompromiss zwischen den Zielen der „richtige" ist, wird die Person eine sicherheitstechnische Barriere sogar aktiv abbauen. Bekanntestes Beispiel hierzu ist der Eingriff in den Reaktorschutz durch die Betriebsmannschaft von Tschernobyl, um einen Test durchzuführen.

Dabei geht es nicht allein um die tatsächlich am technischen System handelnde Person oder Personengruppe. Bei schweren technischen Störungen sind insbesondere Zielkonflikte aus organisatorischen Überlegungen heraus eine wesentliche Ursache für das Zustandekommen eines Unfalls. Die wesentlichen organisatorischen Gründe für Zielkonflikte können sein (Hollnagel 2009):

- Zielkonflikte auf der Organisationsebene hinsichtlich Kosten- und Nutzenbetrachtungen
- Zielkonflikte auf der Managementebene (z. B. Abwägungen des Mitarbeiters, bei einem nur beinahe kritischen Befund eine Handlung mit drastischen Konsequenzen durchzuführen)
- Zielkonflikte auf der Arbeits- oder Instandhaltungsebene zwischen Durchführung einer monotonen Aufgabe und Ablenkung durch andere Tätigkeiten

Bezogen auf Entsorgungsoptionen, in denen der Mensch mit langwierigen, trägen physikalischen Phänomenen zu tun hat, kann auf Basis dieser Überlegungen kein aufgabenbezogenes Verhalten in der Entsorgung angenommen werden. Dass ein Mensch mit gleichbleibender Zuverlässigkeit ein Endlager, in dem keine unmittelbaren täglichen aufgabenbezogenen Handlungen erforderlich sind, über Jahrzehnte/Jahrhunderte aufgabenbezogen kontrollieren und überprüfen wird, kann ausgeschlossen werden.

Eine nachhaltig sichere Entsorgungsoption benötigt deshalb ein Sicherheitskonzept, das den zielbezogenen Aspekt von Mensch und Organisation berücksichtigt (Hollnagel et al. 2005). Aus der Problematik der Zielkonflikte ergibt sich, dass Kern jeglicher nachhaltig sicheren Entsorgungsoption ein vernünftiger Umgang mit den Zielkonflikten im Entsorgungssystem sein muss. Der zu betrachtende Systemumfang ergibt sich aus den Arbeitsebenen der Gestaltung.

2.2 Arbeitsebenen der Gestaltung

Ein angemessenes Sicherheitskonzept hinsichtlich Mensch und Organisation muss zunächst eine umfassende Systemsicht einnehmen und nicht allein den operationellen Betrieb eines Endlagers betrachten. Eine klassische Unterteilung der Arbeitsebene bildet das berühmte Schweizer-Käse-Modell, das Reason (1997) aufgestellt hat und das von Leveson (2002) und Hollnagel (2004) weiterentwickelt wurde (zusammenfassend Abb. 1).

Abbildung 1 zeigt, welche Arbeitsebenen für die nachhaltige Gestaltung eines Systems zu berücksichtigen sind. Der Systembegriff umfasst die sogenannte Ausführungsebene – das sind all die Personen und Personengruppen, die tatsächlich in dem technischen System agieren – bis hin zur gesellschaftlichen bzw. regelgebenden Ebene. Dazwischen befinden sich weitere Ebenen, die die Gesamtleistung des Systems bestimmen: die interorganisationale Ebene, die Organisationsebene, die Management- und die Instandhaltungsebene.

Abbildung 1: Arbeitsebenen eines Gesamtsystems in der Unfallenstehung

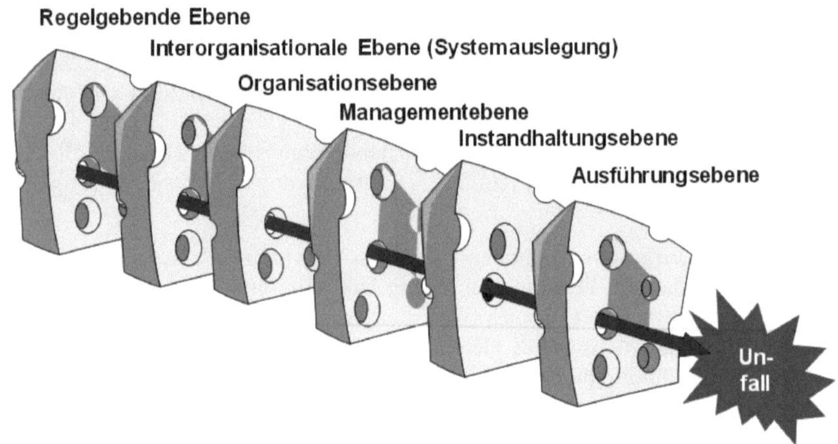

Quelle: Eigene Darstellung

Zielorientiertes Verhalten ist auf allen Ebenen wirksam; langfristige Auswirkungen hat insbesondere zielorientiertes Verhalten auf der regelgebenden und der interorganisationalen Ebene, da diese die Systemauslegung und die gesellschaftlichen Anforderungen festlegen, träge reagieren und bei Bedarf langwieriger Prozesse der Optimierung bedürfen.

Um eine nachhaltige Entsorgungsoption zu betreiben, sind alle Ebenen hinsichtlich ihres zielbezogenen Verhaltens zu untersuchen und gerade die sich zwischen den Ebenen entwickelnden Zielkonflikte zu gestalten, wenn eine Entsorgungsoption nachhaltig sicher betrieben werden soll. Ein typisches, zwischen den Ebenen bestehendes Problem wäre der Zielkonflikt zwischen einer politischen Anforderung und der Möglichkeit einer Organisation, diese umzusetzen. Diese Zielkonflikte müssen dabei nicht notwendigerweise gelöst werden (und können teilweise auch gar nicht gelöst werden); sie müssen nur in der Gestaltung des Systems so berücksichtigt werden, dass sie nicht zu sicherheitstechnisch unerwünschten Zuständen führen. Werden diese Zielkonflikte nicht gestaltet, wird ein Prozess in Gang gesetzt, der als „drift into failure“ bezeichnet wird.

2.3 Drift into failure

Das Grundprinzip des sicheren Betriebs einer Entsorgungsoption ist das permanente Hinterfragen aller Arbeitsebenen hinsichtlich der Sicherheitsleistung und des Sicherheitsbeitrages. Ein wichtiger Aspekt in der Entsorgung ist das langfristige Aufrechterhalten des Hinterfragens. Woods (2003) hat eine geeignete Sichtweise auf das Gesamtsystem in der Fehlerforschung entwickelt, die als „drift into failure“ bekannt geworden ist. Er hat Großereignisse technischer Art, also Ereignisse wie Fukushima oder Tschernobyl oder auch in der Öl- und Gasindustrie oder dem Luftfahrtbereich untersucht und eine typische Entwicklung von Fehlern über die verschiedenen Arbeitsebenen festgestellt. Daraus hat er Systematiken entwickelt, was ein hinterfragendes System abfangen muss (Abb. 2).

Abbildung 2: Drift into Failure

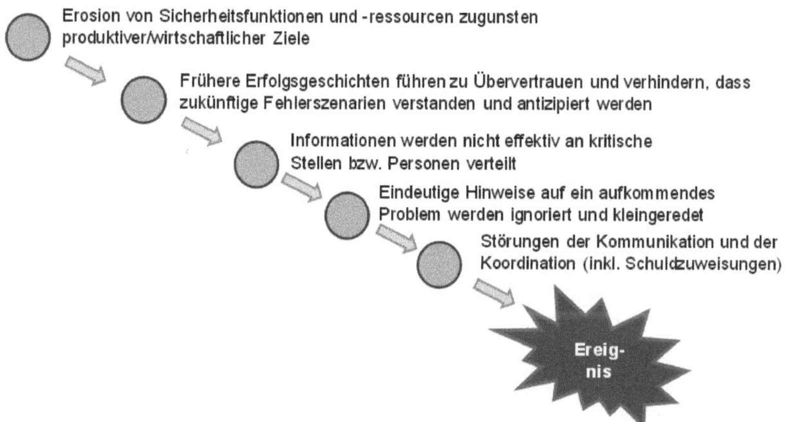

Quelle: Eigene Darstellung

Eine negative Entwicklung hinsichtlich der langfristigen Sicherheitsziele beginnt eigentlich immer mit der Degradierung von Sicherheitsfunktionen zugunsten irgendwelcher alternativen Ziele. Das heißt, dass sich die Gesellschaft oder eine Organisation in irgendeiner Form entschieden hat, Sicherheitsfunktionen nicht mehr so intensiv durchzuführen, wie es vielleicht ursprünglich der Fall war. Zugunsten von Produktivität oder anderen Aspekten wird die Entscheidung getroffen, bestimmte Dinge nicht wie bisher oder wie es sicherheitstechnisch sinnvoll wäre, durchzuführen. Ein berühmtes Beispiel hierfür ist der BP-Unfall im Golf von Mexiko – hier wurden tatsächlich Sicherheitsfunktionen aufgrund von wirtschaftlichen Zielstellungen vernachlässigt (DHSG 2011). Andere bekannte Beispiele sind das Auslassen von Wartungs- oder Prüfintervallen zugunsten der Wirtschaftlichkeit im Flugbetrieb.

Der zweite Aspekt beinhaltet, sich auf getroffenen Entscheidungen oder Erfolgsgeschichten auszuruhen, statt diese Zeit der Sicherheit zu nutzen, um zukünftige ggf. nachteilig wirkende Entwicklungen zu antizipieren. Dieser zweite Aspekt ist als einer der wichtigsten für eine Entsorgungsoption anzusehen, denn egal, wie diese letztendlich im Detail aussieht, fängt damit im Sinne der Fehlerforschung die eigentliche Hinterfragung des Systems an: Ist das der richtige Weg? Wie müssen bestimmte politische/gesellschaftliche/technologische Entwicklungen berücksichtigt werden, um in den kommenden Jahren sicher zu sein?

Eine der Hauptaufgaben für eine Organisation, die die Entsorgungsoption betreibt, besteht demzufolge darin, fortlaufende strategische Planungen hinsichtlich aller Arbeitsebenen für die kommenden Jahre und Jahrzehnte zu betreiben, um nachhaltig agieren zu können.

Der dritte Punkt des Konzepts zeigt die Reaktanz und Trägheit des Gesamtsystems auf, rechtzeitig auf Probleme zu reagieren. Liegen Sicherheitsprobleme vor, wird oft sehr spät realisiert, dass etwas zu tun ist. Denn sobald ein solcher Entscheidungspfad eingeschlagen wurde, wird keine Überprüfung dieses Pfades mehr vorgenommen und trotz bereits eindeutiger Hinweise wird nicht rechtzeitig agiert. Es findet keine systematische Überwachung mehr statt, und es wird keine Handlungsoption entwickelt oder diese wird zu spät entwickelt. In den Spätphasen eines Ereignisses ist zudem eine mangelnde Koordination zwischen den entsprechenden Stellen oder ein Denken in starren Strukturen zu verzeichnen.

Das Konzept des „drift into failure" beschreibt langfristige Entwicklungsaspekte, die sich in Zeiträumen von Dekaden abspielen. Eine politische oder wirtschaftliche Entscheidung wirkt sich also nicht direkt, sondern üblicherweise erst nach Dekaden auf der Arbeitsebene aus. Entsprechend müssen die Prozesse von den ersten Entscheidungen auf politischer Ebene bis zur Arbeitsebene über ein hinterfragendes System abgefangen werden. Um dies zu erreichen, ist eine hohe Sicherheitskultur und eine systematische Überprüfung von Zielkonflikten auf den oberen Arbeitsebenen erforderlich; diese kennzeichnen eine hinterfragende Organisation.

3 Kennzeichen hinterfragender Organisationen

3.1 Sicherheitskultur

Der bekannteste Ansatz zur Herstellung nachhaltiger hinterfragender Organisationen ist das Konzept der Sicherheitskultur. Sicherheitskultur wird üblicherweise umschrieben als eine sicherheitsgerichtete Grundhaltung, Verantwortung und Handlungsweise aller Mitarbeiter auf allen Hierarchiestufen, die sich in sicherheitsgerichteten Tätigkeiten äußert und über die erforderlichen Tätigkeiten zur Erfüllung interner oder externer Anforderungen hinausgeht. Sicherheitskultur umfasst die Gesamtheit der Eigenschaften und Verhaltensweisen innerhalb eines Unternehmens und beim Einzelnen (Balfanz et al. 2004).

Sicherheitskultur umfasst primär die Organisation und ihre Mitglieder und hat zum Ziel, sicherheitstechnisch ungünstiges zielorientiertes Verhalten auf allen Ebenen der Organisation, also von der Organisationsebene bis hin zur Ausführungsebene, zu verhindern. Unberücksichtigt bleiben in dem Konzept die externen Einflüsse auf die Organisation aus der interorganisationalen Ebene oder der regelgebenden/politischen Ebene. Deshalb greift das Konzept für eine Organisation der Endlagerung zu kurz, denn der Einfluss dieser Ebenen wird zu zielorientiertem Verhalten auf der Organisationsebene führen und damit Aktivitäten hinsichtlich der Sicherheitskultur in der Organisation neutralisieren (Patterson 2002).

Eine Sicherheitskultur über alle Arbeitsebenen des Gesamtsystems hinweg wäre dagegen eine sinnvolle Sichtweise; diese kann als interorganisationale Sicherheitskultur verstanden werden.

3.2 Reifegrad des Gesamtsystems

Das Gesamtsystem aller Arbeitsebenen kann mithilfe von Reifegraden bewertet werden. Bezogen auf die Art, wie ein System agieren sollte, wird zwischen pathologischen Systemen im schlechtesten Fall und generativen Systemen im besten Fall unterschieden (Reason 1997). Dazwischen gibt es eine Reihe von Abstufungen von reaktivem, berechnendem und proaktivem Agieren. Ziel eines hinterfragenden Systems sollte sein, ein generatives System zu etablieren, also nicht in irgendeiner Form durch feste, starre Strukturen zu agieren, sondern diese kontinuierlich zu hinterfragen. Das Faktum der Gewährleistung der Sicherheit sollte zum Kern des Betriebes gemacht werden, um entsprechend erfolgreich zu sein und nicht in irgendwelche Probleme hineinzugeraten (vgl. Tab. 1).

Viele Systeme sind aufgrund von Vermeidungsstrategien auf den unterschiedlichen Arbeitsebenen pathologisch organisiert. Ein typisches Beispiel dafür ist eine Organisation, die in der Kommunikation der zentralen Firmenpolitik (einer Vision oder Mission der Organisation) der Sicherheit höchste Priorität einräumt, aber im tatsächlichen Handeln wirtschaftlichen Interessen den Vorrang gibt. Mitarbeiter sind dann gezwungen, den impliziten Zielkonflikt zwischen der zentralen Firmenpolitik und den impliziten Managementerwartungen zu lösen. Das Resultat sind Ambiguitäten und Vermeidungsverhalten. In solchen Fällen geht die Fehlerforschung davon aus, dass früher oder später Sicherheitsprobleme entstehen.

Tabelle 1: Reifegrad eines Gesamtsystems

Reifegrad	*Kennzeichnung/Beschreibung*
	Die Organisation ...
Generativ	... legt permanent einen Fokus auf Sicherheit ... sieht Sicherheit als Wettbewerbsfaktor, nicht als Kostenfaktor ... nimmt neue Ideen der Belegschaft und des Organisationsumfeldes zum Thema Sicherheit aktiv auf
Proaktiv	... plant Ressourcen ein, um proaktiv Unfallvermeidung zu betreiben ... wird sowohl offen und flexibel als auch mittels gut geführter Statistiken informiert und geleitet ... sorgt dafür, dass die Belegschaft die Arbeitsverfahren mitträgt und von sich aus weiterentwickelt
Berechnend	... legt Sicherheit als Unternehmensziel fest ... verfolgt intensive Audits zur Reduktion von Unsicherheit ... führt Indikatoren ein und generiert umfangreiche Statistiken zum Thema Sicherheit
Reaktiv	... stellt normatives Verhalten als einzig seligmachendes Verhalten in den Vordergrund ... beschönigt oder „uminterpretiert" Ereignisse ... schenkt Sicherheit nach einem Unfall eine gewisse Aufmerksamkeit
Pathologisch	... begnügt sich mit der Einhaltung der Gesetze ... akzeptiert, dass Unfälle passieren ... sucht den Verantwortlichen bzw. Schuldigen

Quelle: Eigene Darstellung

3.3 Monitoring und Safety-Scanning

Für die unterschiedlichen Reifegradstufen gibt es entsprechende Monitoring-Verfahren, um den Zustand einer Organisation oder eines Gesamtsystems – bezogen auf diese Reifegrade – regelmäßig zu untersuchen und diese entsprechend zu entwickeln. Vier Verfahrensgruppen können dabei unterschieden werden, die wiederum auf allen Arbeitsebenen zu berücksichtigen wären:

(1) Individuelle kognitive Aspekte der Wahrnehmung und des Umgangs mit den Limitationen der menschlichen Informationsverarbeitung.
(2) Organisatorische Aspekte zwischen den Ebenen hinsichtlich des Informationsflusses und der Transparenz im Falle von sicherheitskritischen Auseinandersetzungen.
(3) Normative oder rechtliche Aspekte, welche alle rechtlichen und normativen Setzungen im Gesamtsystem umfassen und die so gestaltet sein müssen, dass insgesamt die Interaktion zwischen den Ebenen nicht beeinträchtigt wird.
(4) Kybernetische Aspekte im Zusammenspiel aller Elemente des Gesamtsystems (technische und menschliche).

Zu diesen vier Aspekten existiert eine Reihe von Methoden zum Aufbau selbsthinterfragender Systeme. Bezogen auf die individuellen kognitiven Aspekte ist dies beispielsweise das Workload-Management, mit dem die Personen bezogen auf ihre Selbsthinterfragung geschult werden. Bezogen auf die Prozesse innerhalb einer Organisation ist die wohl bekannteste Methode das sogenannte Crew Ressource Management (Helmreich/Merritt 1998; Helmreich/Sexton 2002). Es stammt aus der Luftfahrt und wird insbesondere für Piloten genutzt, um deren Teamverhalten sicherzustellen, wurde aber für vielfältige andere technische Domänen weiterentwickelt (Dietrich/Childress 2004). Auf der Ebene der Systemgestaltung greift das sogenannte Safety Scanning, das zum Monitoring und zur Entscheidungsfindung dienen kann (EUROCONTROL 2011; Sträter et al. 2012). Abbildung 3 zeigt die Gesamtstruktur des Safety Scanning.

Das Safety Scanning moderiert das kritische Hinterfragen von regulatorischen Aspekten (Aufsicht, juristische Aspekte), interorganisatorischen (Systemgestaltung), organisatorischen sowie operativen Aspekten. Es bildet das Gesamtsystem ab und hinterfragt Beiträge und Wechselwirkungen zwischen den Ebenen. Es vermeidet zielorientiertes Verhalten auf allen Arbeitsebenen, indem es bereits in frühen strategischen Entwicklungsphasen alle Interessenvertreter und Arbeitsebenen in systematischer Art und Weise zusammenbringt und Zielkonflikte miteinander löst (EUROCONTROL 2011; Korteweg et al. 2012).

Abbildung 3: Gesamtstruktur des Safety Scanning

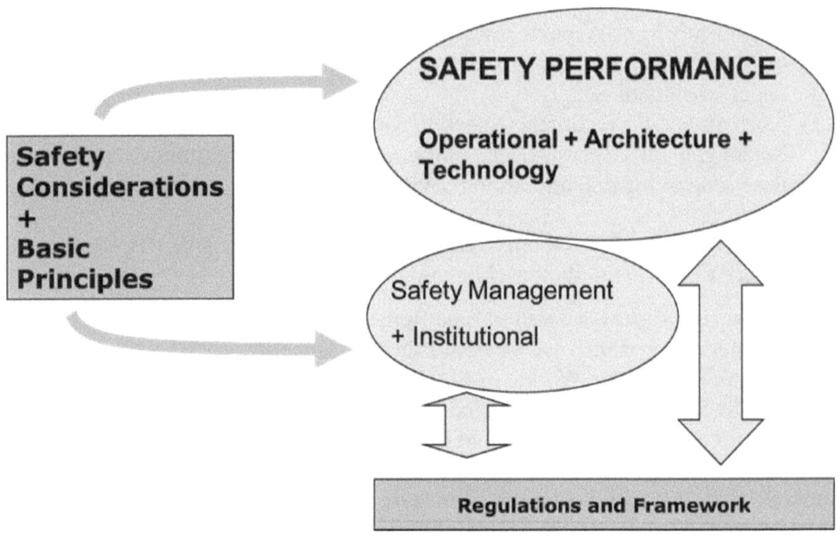

Quelle: EUROCONTROL 2011

4 Diskussion

Die vorhergehenden Abschnitte haben die Aspekte einer nachhaltigen Gestaltung des Gesamtsystems Endlagerung aus menschlicher und organisatorischer Sicht umrissen. Die Frage nach den Akteuren und deren Gestaltung des Systems soll abschließend diskutiert werden.

4.1 Moderator des Prozesses im Gesamtsystem

Irgendeine Organisation muss den Prozess der nachhaltigen Sicherheitsgestaltung in der Hand haben; sie muss sich um die operativen Aspekte der unteren Arbeitsebenen genauso kümmern, wie um die richtige Gestaltung des interorganisationalen und gesellschaftlichen Umfeldes.

Nur wenn beides gelingt und die Organisation vernünftig in das Gesamtsystem eingebunden ist, kann ein sicherer Betrieb über einen sehr langen Zeitraum aufrechterhalten werden.

Diese Organisation muss zunächst einen systematischen Ansatz zum Monitoring aufbauen, in dem alle Arbeitsebenen angesprochen und moderiert werden. Der Monitoring-Prozess sollte so neutral wie möglich gestaltet sein, d. h. er sollte nicht durch Zielkonflikte belastet sein bzw. nicht bereits im Vorhinein unlösbare Zielkonflikte einseitig festgelegt haben. Hierzu kann ein Ansatz auf Basis des Safety Scanning dienen. Über die fundamentalen Sicherheitsanforderungen werden zielorientierte Interessen der unterschiedlichen Arbeitsebenen ausgeglichen und die zukünftigen Anforderungen an die Entsorgungsoption entwickelt. Das Verfahren muss iterativ und regelmäßig den Sicherheitszustand des Gesamtsystems erfassen und bewerten und rechtzeitig auf der strategischen Planungsebene Weichen stellen, damit über den gesamten Zeitraum der Entsorgungsoption ein gleichbleibendes Sicherheitsniveau aufrechterhalten wird.

4.2 Rolle der Aufsicht im Gesamtsystem

Eine Entsorgungsoption mit entsprechend langem Zeithorizont erfordert ein Umdenken hinsichtlich der Rolle der Aufsicht. Üblicherweise wird eine juristische und strenge Trennung zwischen Aufsicht und Betrieb festgelegt. Das ist für eine kontinuierliche strategische Entwicklung unvorteilhaft, denn durch die strenge Trennung entstehen retardierende Elemente im Gesamtsystem, insbesondere zwischen Betreiberorganisation und regelgebender Ebene. Der Grund liegt darin, dass die Aufsichtsstruktur sich oft als normative Kraft sieht und als die richtige Richtung vorgebende Institution. Diese klassische Sicht einer Aufsichtsbehörde ist bei einem Endlager sicher sinnvoll für die operativen Ebenen des Betreibers; allerdings sind die strategischen Fragestellungen der kontinuierlichen strategischen Planung nicht in einem klassischen Aufsichtsverfahren durchführbar. Die Aufsichtsstruktur muss aktiv hinterfragt werden dürfen und sich den gesellschaftlichen Veränderungen und den Bedürfnissen der Sicherheit der Endlageroption anpassen.

Die Aufsichtsbehörde ist ein Teilsystem, das sich selbst permanent hinterfragen muss. Das Safety Scanning sieht dafür z. B. einen Abschnitt zur Hinterfragung der Aufsichtsführung vor. Die Aufsichtsbehörde muss sich in einer langfristig sicheren Endlageroption von der Rollendefinition her nicht als allwissende normative Macht verstehen, sondern sich als Teil eines hinterfragenden Systems sehen.

Damit verbunden ist auch ein zentrales rechtliches Problem im Rahmen eines solchen hinterfragenden Systems, denn letztendlich kann nur das Gesamtsystem und nicht allein die Betreiberorganisation für mögliche Schäden haften. Als Ausweg kann die Aufsichtsbehörde als Treiber für das aktive Hinterfragen gesehen und eine entsprechende Aufsichtskultur entwickelt werden.

4.3 Interaktion im Gesamtsystem

Bezogen auf die Interaktion im Gesamtsystem muss eine kritikfreundliche Grundstimmung erzeugt werden, wenn das Gesamtsystem nachhaltig sicher sein soll. Hierzu gehört eine Kultur der Kritik, bestehend aus einer Qualifikation der Kritikfähigkeit und Kritikvermittlung, einer entsprechenden interorganisationalen Umgangskultur und auch etablierter Prozesse zum Umgang mit Kritik.

Dies gilt auch für den Umgang mit Whistleblowern: Da sie nicht in das Gesamtbild hineinpassen, werden Kritik und Kritiker üblicherweise aus dem System herausgenommen. Dadurch entsteht für die Organisation zunächst scheinbar ein Erfolg (der Störenfried ist weg, alles kann ungestört weiterlaufen). Das Problem, das der Kritiker nannte, ist aber noch vorhanden und kann ungehindert wirken. Die Organisation verliert jedoch ausgerechnet den Kompetenzträger, der das Problem zu lösen vermag (Sträter et al. 2013). Sicherheitskritische Zustände und schwere Unfälle sind das Resultat dieses Vorgehens. Das berühmteste Beispiel dafür ist sicher die Challenger-Katastrophe, in der durch Versagen von Dichtringen die Raumfähre beim Start explodierte. Das Problem war einem Entwicklungsingenieur bekannt, seine Warnungen wurden aber ignoriert (Leveson 2008).

Im Gesamtsystem muss eine entsprechende Fehlerkultur etabliert werden. Sie gehört mit zur Qualifizierung des Gesamtsystems. Fehlerkultur bedeutet zum Beispiel, Fehlentscheidungen zuzugeben, den Fehler nicht als persönlichen Makel hinzustellen, sondern als Hinweis dafür, dass ein Problem gelöst werden oder eine ehemals für sinnvoll erachtete Lösung in irgendeiner Form revidiert werden muss. Dieser Aspekt der Fehlerkultur ist ein begleitender Aspekt für alle Personen, die im engeren Kreis im Gesamtsystem interagieren (siehe hierzu z. B. Woods 2003).

Die Effekte inadäquater Berücksichtigung psychologischer Aspekte in Organisationsprozessen können auf unterschiedlichen Ebenen definiert werden:

- Auf der individuellen Ebene sind insbesondere mangelnde Kreativität und Motivation oder innere Kündigung bekannte psychologische Auswirkungen, d. h., Wissen über Sicherheitsprobleme wird nicht mehr geteilt.
- Auf der betriebswirtschaftlichen bzw. organisationalen Ebene können sich diese in Qualitätseinbußen, mangelnder Flexibilität der Organisationen, auf Änderungen zu reagieren, äußern, d. h., die Organisation erkennt den Handlungsbedarf zu spät.
- Auf der volkswirtschaftlichen bzw. gesellschaftlichen Ebene ist davon auszugehen, dass sich diese negativ auf die Wettbewerbsfähigkeit und Anpassungsfähigkeit einer Gesellschaft auswirken, d. h., es entstehen u. U.

Generationenkonflikte und hierüber mangelnde Sorgfalt im Umgang mit der Endlageroption.

Reason hat dieses Phänomen mit dem Satz umschrieben „If you think you are safe you did your first mistake"; wenn man Kritiker außen vor lässt, dann hat man eigentlich schon den ersten Fehler in Richtung „drift into failure" gemacht. Das heißt, Kritik muss im Gesamtsystem, systematisch in den Arbeitsebenen und zwischen den Arbeitsebenen fließen können. Kritik muss als positive Information zur Verbesserung des Gesamtsystems verstanden werden. Ziele müssen sein: die Früherkennung und positive Aufnahme von Kritik als Wissen, die Schaffung von Gestaltungsfreiräumen und die Vermeidung von Überkontrolle (und damit von Misstrauenskultur) sowie die positive Beeinflussung der Motivation der Mitarbeiter durch eine gute Betriebs- und Führungskultur. Abbildung 4 fasst hierzu essenzielle Anforderungen an die Interaktion zusammen.

Abbildung 4: Essenzielle Anforderungen an die Interaktion im Gesamtsystem

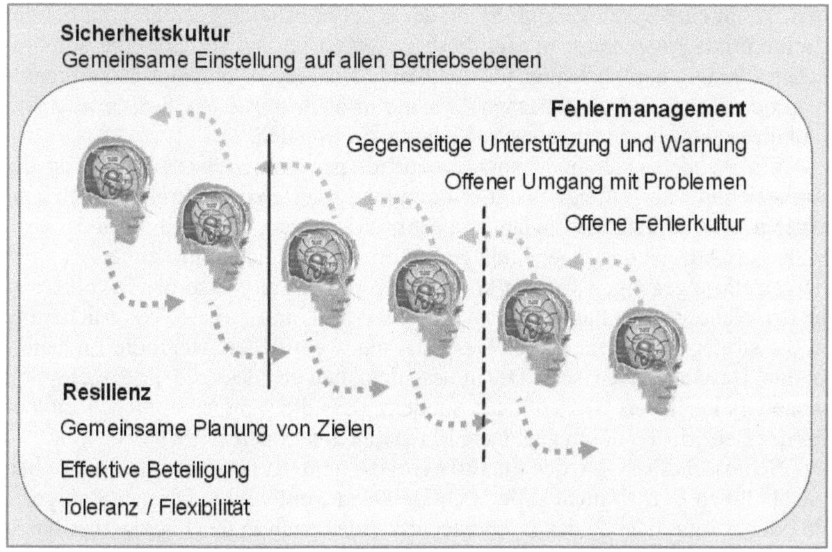

Quelle: Eigene Darstellung

Um einen solchen Gesamtprozess aufzubauen, bei dem auch die Stakeholder nicht genau bekannt sind, muss die Planung als ein iteratives Verfahren ausgelegt werden. Das heißt, man würde nicht einmalig in irgendeiner Form einen Entwicklungsprozess starten und diesen dann als „closed shop" weiterverfolgen. Die Gruppe der entsprechenden Wissensträger muss über den gesamten Prozess offen sein und Wissensträger müssen nach einem systematischen Verfahren berücksichtigt werden.

Das gilt auch für das Monitoring, in dem systematisch und regelmäßig der Verlauf der Entwicklung nach denselben Prinzipien überprüft werden muss, also z. B. das Safety Scanning iterativ eingesetzt wird, bis das Bild aller Stakeholder vollständig ist und eine strategische Planung erfolgversprechend sein kann.

5 Ausblick

Um einen psychologisch nachhaltigen und sicheren Betrieb einer Endlageroption aufzustellen, ist die Etablierung eines selbsthinterfragenden Gesamtsystems entscheidend. Dies bedingt eine Offenheit aller beteiligten Parteien und ein systematisches Verfahren, auch konfligierende Ziele miteinander anzugehen. Wenn eine Systematik im Sinne des Safety Scanning genutzt wird, bei der Zielkonflikte mitgenommen und diese systematisch und seriös in der strategischen Planung berücksichtigt werden, dann setzt ein psychologischer Erkenntnisprozess ein, und es entstehen Lösungsansätze, mit Zielopfern umzugehen und diese aktiv in die strategische Planung einzubinden.

Ein weiteres, hier nicht angesprochenes psychologisches Moment ist die Einbeziehung der Öffentlichkeit, Kommunen oder Expertenkreise. Auch hier ergeben sich zusätzliche fundamentale Anforderungen, die im Rahmen der strategischen Ausplanung einer Endlageroption nach ähnlichen Prinzipien wie hier dargestellt angegangen werden können. Wenn beispielsweise die Bevölkerung in der Nähe der Endlageroption verunsichert ist, muss diese Verunsicherung in die strategische Planung des Gesamtsystems einfließen, wenn die Endlageroption Bestand haben soll. Das muss nicht heißen, dass die Endlageroption woanders lokalisiert werden muss, sondern, dass die wahrgenommenen Zielopfer der Betroffenen durch geeignete Maßnahmen kompensiert werden.

Schlussendlich soll der Qualifizierungs- und Kompetenzerhalt für einen nachhaltigen Betrieb nicht außer Acht gelassen werden. Wie können Studierende oder Auszubildende davon überzeugt werden, sich in der Endlageroption zu engagieren? Wie können die damit verbundenen Fragestellungen in Studien- oder Ausbildungsgänge eingebunden werden, sodass eine zukunftsorientierte Perspektive für das zukünftige Personal eines Endlagers entsteht. Das bedingt auch eine entsprechende Anpassung der Ausbildungsstrukturen, die im Gesamtsystem der interorganisationalen Ebene zuzuordnen sind.

Literatur

Apostolakis, G.; Soares, C.G.; Kondo, S.; Sträter, O. (Hg.) (2004): Human Reliability Analysis: Data Issues and Errors of Commission. Special Edition. Reliability Engineering & System Safety 83(2), S. 127–276

Balfanz, H.; Junge, R.; Linsenmaier, B.; Rausch, H.; Sträter, O.; Kallenbach-Herbert, B.; Sauerbrey, U.; Sickert, M.; Oehmgen, T.; Rotzsche, J. (2004): Entwicklung praxisgerechter Bewertungskriterien für die Sicherheitskultur in deutschen Kernkraftwerken. Hamburg

DHSG – Deepwater Horizon Study Group (2011): Final Report on the Investigation of the Macondo Well Blowout. Berkeley

Diedrich, R.; Childress, T.M. (Hg.) (2004): Group Interaction in High Risk Environments. Aldershot

EUROCONTROL (2011): SRC Document 46 – Safety Scanning. EUROCONTROL Safety Regulation Commission. Brüssel

Helmreich, R.L.; Merritt A.C. (1998): Culture at Work in Aviation and Medicine: National, Organizational, and Professional Influences. Aldershot

Helmreich, R.L.; Sexton, J.B. (2002): Managing threat and error to increase safety in medicine. 6. Berliner Kolloquium: Interaktion unter Hoch-Risiko-Bedingungen. 6. Mai 2002. Gottlieb Daimler- and Karl Benz-Stiftung, Berlin

Hollnagel, E. (2004): Barriers and Accident Prevention. Aldershot

Hollnagel, E. (2009): The ETTO Principle: Efficiency-Thoroughness Trade-Off: Why Things That Go Right Sometimes Go Wrong. Aldershot

Hollnagel, E.; Woods, D.; Leveson, N. (2005): Resilience Engineering – Concepts and Precepts. Aldershot

Kahneman, D.; Tversky, A. (1979): Prospect Theory: An Analysis of Decision Under Risk. Ecometrica 47, S. 263–291

Korteweg, H.; Straeter, O.; Athanassiou, G. (2012): Safety Scanning for the Single European Sky. In: Gesellschaft für Arbeitswissenschaft (Hg.): Gestaltung nachhaltiger Arbeitssysteme – Wege zur gesunden, effizienten und sicheren Arbeit. Dortmund, S. 65–68

Leveson, N. (2002): System Safety Engineering: Back to the Future. Boston, MA

Leveson, N. (2008): Technical and Managerial Factors in the NASA Challenger and Columbia Losses: Looking Forward to the Future In: Kleinman, D.L.; Cloud-Hansen, K.A.; Matta, C.; Handelsman, J. (Hg.): Controveries in Science and Technology. Volume 2. New Rochelle, NY

Mosneron-Dupin, F.; Reer, B.; Heslinga, G.; Sträter, O.; Gerdes, V.; Saliou, G.; Ullwer, W. (1997): Human-Centered Modeling in Human Reliability Analysis: Some Trends Based on Case Studies. Reliability Engineering and System Safety 58(3), S. 249–274

OECD (2004): Technical Opinion Papers No. 4 – Human Reliability Analysis in Probabilistic Safety Assessment for Nuclear Power Plants. OECD NEA No. 5068. Paris

Patterson, I. (2002): Organizational Culture and Safety Culture as Determinants of Error and Safety Levels in Aviation Maintenance Organizations: A Latent Failure Approach Thesis. Albany, New Zealand

Reason, J. (1990): Human Error. Cambridge

Reason, J. (1997): Managing the Risk of Organizational Accidents. Aldershot

Sträter, O. (2005): Cognition and Safety – An Integrated Approach to Systems Design and Performance Assessment. Aldershot

Sträter, O.; Korteweg, H.; Nollet, J.; Everdij, M.; Athanassiou, G.; Arenius, M.; Kraan, B. (2012) Safety Scanning – An Approach to Manage Safety in the Single European Sky. Paper presented at the PSAM/ESREL Conference 2012, Helsinki

Sträter, O.; Reer, B.; Dang, V.; Hirschberg, S. (1999): Methods, Case Studies, and Prospects for an Integrated Approach for Analyzing Errors of Commission. In: Schueller, G.; Kafka, P. (Hg.): Safety and Reliability. Rotterdam. S. 699

Sträter, O.; Siebert-Adzic, M.; Schäfer, E. (2013): Gesundes Führen für effiziente Organisationen der Zukunft. In: Grote, S. (Hg.): Die Zukunft der Führung. Heidelberg, S. 307–330

Swain, A.D. (1989): Comparative Evaluation of Methods for Human Reliability Analysis. GRS-71, Köln

Swain, A.D.; Guttmann, H.E. (1983): Handbook of Human Reliability Analysis with Emphasis on Nuclear Power Plant Applications. NUREG/CR-1278, Washington, DC

VDI 4006 Blatt 2 (2017): Menschliche Zuverlässigkeit. Methoden zur quantitativen Bewertung menschlicher Zuverlässigkeit. Berlin

Woods, D. (2003): Creating Foresight: How Resilience Engineering Can Transform NASA's Approach to Risky Decision Making. Testimony on The Future of NASA for Committee on Commerce, Science and Transportation. Washington, DC

Autorenverzeichnis

Stefan Böschen, Dr. phil.: Universitätsprofessor, Lehrstuhl für Technik und Gesellschaft an der RWTH Aachen, Theaterplatz 14, D-52062 Aachen, E-Mail: stefan.boeschen@humtec.rwth-aachen.de

Anne Eckhardt, Dr. sc. nat. ETH: risicare GmbH, Bühlstrasse 19, CH-8125 Zollikerberg, E-Mail: anne.eckhardt@risicare.ch

Ida Epkenhans, MSc., Univ.-Prof. Dr.-Ing.: Institut für Geomechanik und Geotechnik der TU Braunschweig, Beethovenstr. 51b, D-38106 Braunschweig, E-Mail: i.epkenhans@tu-braunschweig.de

Armin Grunwald, Prof. Dr.: Institutsleiter am Institut für Technikfolgenabschätzung und Systemanalyse (ITAS), Karlsruher Institut für Technologie (KIT), Leiter des Büros für Technikfolgen-Abschätzung beim Deutschen Bundestag (TAB), Professor für Technikphilosophie am Institut für Philosophie des KIT, Karlstraße 11, D-76133 Karlsruhe, E-Mail: armin.grunwald@kit.edu

Thomas Hassel, Dr.-Ing.: Bereichsleiter Unterwassertechnikum Hannover (UWTH) am Institut für Werkstoffkunde (IW) der Leibniz Universität Hannover (LUH), Lise-Meitner-Str. 1, D-30823 Garbsen, E-Mail: hassel@iw.uni-han nover.de

Peter Hocke, Dr. phil.: Forschungsgruppe „Endlagerung als soziotechnisches Projekt“ am Institut für Technikfolgenabschätzung und Systemanalyse (ITAS), Karlsruher Institut für Technologie (KIT), Karlstraße 11, D-76133 Karlsruhe, E-Mail: hocke@kit.edu

Michael Jobmann, Dipl.-Geophysiker: Abteilung „Endlagertechnik“ in der BGE TECHNOLOGY GmbH, Eschenstraße 55, D-31224 Peine, E-Mail: michael.jobmann@bge.de

Ansgar Köhler, Dipl.-Ing., heute: Fachgruppe Transport und Handhabung, TÜV NORD EnSys GmbH & Co. KG, Am TÜV 1, D-30519 Hannover, E-Mail: anskoehler@tuev-nord.de

Sophie Kuppler, Dr. rer. pol.: Forschungsgruppe „Endlagerung als soziotechnisches Projekt“ am Institut für Technikfolgenabschätzung und Systemanalyse (ITAS), Karlsruher Institut für Technologie (KIT), Karlstraße 11, D-76133 Karlsruhe, E-Mail: sophie.kuppler@kit.edu

Rocio Paola León Vargas, MSc. Dipl.-Ing., heute: BGE TECHNOLOGY GmbH, Eschenstraße 55, D-31224 Peine, E-Mail: Paola.LeonVargas@bge.de

Anna-Laura Liebenstund, M.A., heute: Leiterin der Geschäftsstelle, NordAllianz Metropolregion München Nord, Schloßstr. 2, D-85737 Ismaning, E-Mail: liebenstund@nordallianz.de

Karl-Heinz Lux, Univ.-Prof. Dr.-Ing. habil.: Lehrstuhl für „Geomechanik und multiphysikalische Systeme“ am Institut für Endlagerforschung (IELF), Technische Universität Clausthal (TUC), Erzstraße 20, D-38678 Clausthal-Zellerfeld, E-Mail: lux@tu-clausthal.de

Melanie Mbah, Dr. rer. nat.: Forschungskoordinatorin für Transdisziplinäre Nachhaltigkeitsforschung am Öko-Institut e. V., Merzhauserstraße 173, D-79111 Freiburg, E-Mail: m.mbah@oeko.de

Volker Mintzlaff, Dipl.-Geol.: Institut für Geomechanik und Geotechnik der TU Braunschweig, Beethovenstr. 51b, D-38106 Braunschweig, E-Mail: v.mintzlaff @tu-braunschweig.de

Franziska Semper, Oberregierungsrätin, heute: Leiterin der Internen Revision Finanzministerium M-V, Schlossstraße 9–11, D-19053 Schwerin, E-Mail: franziska.semper@fm.mv-regierung.de

Ulrich Smeddinck, apl. Prof. (Halle) Dr. jur.: Forschungsgruppe „Endlagerung als soziotechnisches Projekt“ (EndFo) am Institut für Technikfolgenabschätzung und Systemanalyse, Karlsruher Institut für Technologie (KIT), Karlstr. 11, D-76133 Karlsruhe, E-Mail: ulrich.smeddinck@kit.edu

Joachim Stahlmann, Univ.-Prof. Dr.-Ing.: Institut für Geomechanik und Geotechnik der TU Braunschweig, Beethovenstr. 51b, D-38106 Braunschweig, E-Mail: j.stahlmann@tu-braunschweig.de

Oliver Sträter, Univ.-Prof. Dr. phil. habil.: Fachgebiet Arbeits- und Organisationspsychologie der Universität Kassel, Heinrich-Plett-Straße 40, D-34132 Kassel, E-Mail: straeter@uni-kassel.de

Ralf Wolters, Dr.-Ing.: Lehrstuhl für „Geomechanik und multiphysikalische Systeme“ am Institut für Endlagerforschung (IELF), Technische Universität Clausthal (TUC), Erzstraße 20, D-38678 Clausthal-Zellerfeld, E-Mail: ralf.wolt ers@tu-clausthal.de

Juan Zhao, Dr.-Ing.: Lehrstuhl für „Geomechanik und multiphysikalische Systeme“ am Institut für Endlagerforschung (IELF), Technische Universität Clausthal (TUC), Erzstraße 20, D-38678 Clausthal-Zellerfeld, E-Mail: juan.zhao@tu-clausthal.de

Zeitfracht Medien GmbH
Ferdinand-Jühlke-Straße 7
99095 Erfurt, Deutschland
produktsicherheit@kolibri360.de